Dalton's Introduction to Practical Animal Breeding

Other books of interest

FARM MACHINERY
12th Edition
Claude Culpin
0-632-03159-X (pb)
0-632-03158-1 (hb)

FREAM'S PRINCIPLES OF FOOD AND AGRICULTURE
17th Edition
Colin Spedding
0-632-2978-1

PLANNED BEEF PRODUCTION AND MARKETING
David Allen
0-632-02611-1

POULTRY HEALTH AND MANAGEMENT
2nd Edition
David Sainsbury
0-632-02074-1

MECHANISED LIVESTOCK FEEDING
David Bebb
0-632-02035-0

UNDERSTANDING THE DAIRY COW
John Webster
0-632-01889-5

PRACTICAL STUD MANAGEMENT
John Rose & Sarah Pilliner
0-632-02031-8

Dalton's Introduction to Practical Animal Breeding

Third edition

Malcolm B. Willis

BSc (Dunelm), PhD (Edin)
Senior Lecturer in Animal Breeding and Genetics
Faculty of Agriculture
University of Newcastle upon Tyne

OXFORD

Blackwell Scientific Publications

LONDON EDINBURGH BOSTON
MELBOURNE PARIS BERLIN VIENNA

Blackwell Scientific Publications
Editorial offices:
Osney Mead, Oxford OX2 0EL
25 John Street, London WCIN 2BL
23 Ainslie Place, Edinburgh EH3 6AJ
3 Cambridge Center, Cambridge,
    Massachusetts 02142, USA
54 University Street, Carlton
    Victoria 3053, Australia

Other Editorial Offices:
Arnette SA
2, rue Casimir-Delavigne
75006 Paris
France

Blackwell Wissenschaft
Meinekestrasse 4
D-1000 Berlin 15
Germany

Blackwell MZV
Feldgasse 13
A-1238 Wien
Austria

First edition of *An Introduction to Practical
    Animal Breeding* by D. C. Dalton
    published by Granada Publishing Ltd
    1980
Reprinted 1981
Second edition published by Collins
    Professional and Technical Books 1985
Reprinted by BSP Professional Books 1989
Third edition entitled *Dalton's
    Introduction to Practical Animal
    Breeding* published by Blackwell
    Scientific Publications 1991

Set by Excel Typesetters Company Ltd.,
    Hong Kong
Printed and bound in Great Britain by
Hartnolls Ltd, Bodmin, Cornwall

DISTRIBUTORS

Marston Book Services Ltd
PO Box 87
Oxford OX2 0DT
(*Orders*: Tel: 0865 791155
        Fax: 0865 791927
        Telex: 837515)

USA
  Blackwell Scientific Publications, Inc.
  3 Cambridge Center
  Cambridge, MA 02142
  (*Orders*: Tel: 800 759–6102)

Canada
  Oxford University Press
  70 Wynford Drive
  Don Mills
  Ontario M3C 1J9
  (*Orders*: Tel: 416 441–2941)

Australia
  Blackwell Scientific Publications
  (Australia) Pty Ltd
  54 University Street
  Carlton, Victoria 3053
  (*Orders*: Tel: 03 347–0300)

British Library
Cataloguing in Publication Data
Willis, M. B. (Malcolm Beverley)
  Dalton's introduction to practical animal
  breeding. – 3rd. ed.
  1. Livestock. Breeding. Genetic aspects
  I. Title II. Dalton, D. C. (Derek Clive)
  1934–.
  Introduction to practical animal breeding
  636.0821
  ISBN 0–632--03126–3

Library of Congress
Cataloging in Publication Data
Willis, Malcolm Beverley.
  Dalton's introduction to practical
  animal breeding/Malcolm B.
Willis. – 3rd ed.
      p.   cm.
  Includes bibliographical references
  and index.
  ISBN 0–632–03126–3
  1. Livestock – Breeding.   I. Dalton,
  Clive. Introduction to practical animal
  breeding.   II. Title.
SF105.W58 1991
636.08′2–dc20
                                    90–49953
                                         CIP

Contents

Introduction

The term animal breeding refers not so much to the mating, reproduction and rearing of animals, as to the application of the principles of genetics to livestock improvement.

Livestock have been domesticated by man for around 14 000 years, with the dog being probably the first such species followed by sheep, goats, cattle and others. This long association with man does not imply that animal breeding is an ancient practice, although the empirical use of modern principles may well have preceded their scientific definition. There is no doubt that the selection of animals into different forms has been practised over a long period of time. Certain types of dog would appear to have been differentiated into something like their modern forms from biblical times.

The eighteenth century English farmer Robert Bakewell is regarded as the father of modern animal breeding in laying the foundations of the Shire Horse, Longhorn cattle and Leicester sheep. English Thoroughbred horses had their first 'stud book' published in 1791 and Coates' herdbook for Shorthorn cattle appeared in 1822.

Both of these events predated the Austrian monk Gregor Mendel, whose work with the garden pea was formulated in 1865 and was to prove the basis of modern genetics. His work lay largely undiscovered until about 1900. William Bateson, working with chickens, showed in 1901 that Mendelian principles were applicable to animals as well as plants, and modern animal genetics was off and running.

The Hardy–Weinberg law, the basis of population genetics was formulated in 1908, and since that time a great many scientists have contributed to our understanding of animal breeding in one form or another. Early biometricians working in the field were Galton and Karl Pearson while Sir Ronald Fisher and Sewell Wright advanced their ideas to lay the foundations of modern biometrical techniques.

Lush at Iowa, Henderson at Cornell, Robertson at Edinburgh and others such as Van Vleck at Cornell/Minnesota have advanced our understanding of the quantitative and statistical nature of animal breeding, both before and after Watson and Crick had defined the nature of the gene for which they won the Nobel prize in 1953.

Although animal breeding in an empirical sense is quite old it is very young in terms of our scientific understanding of the subject. Most agricultural schools, at whatever level, attempt to teach animal breeding and some principles may also be taught in veterinary schools, although in many cases, alas, not as much as one might wish.

Most people involved with practical breeding of livestock, whether these be farm animals, dogs, cats, guinea pigs or the like, are usually fascinated by the concept of animal breeding, but have only a rough understanding of its principles, especially beyond the Mendelian stage. In contrast, agricultural or veterinary students rarely share this enthusiasm, frequently being put off by the mathematical nature of the subject. Many books have been written seeking to overcome this and to put forward often complex ideas in intelligible form. Falconer's *Introduction to Quantitative Genetics*, first published in 1960, has for 30 years been regarded by many as a standard text for more advanced students to the discipline. Books of a more animal breeding nature and a more introductory level are less obvious. Clive Dalton's *An Introduction to Practical Animal Breeding* (1980) proved very useful for introductory courses at my own University, but since Dr Dalton no longer wishes to continue the revision of that book, this one is offered in its place.

The format follows Dr Dalton's in looking first at the nature of what we wish to do with domestic livestock (Chapter 1). Then follows a brief look at the basic nature of Mendelism (Chapter 2). Population genetics is looked at in Chapters 3 and 4, while Chapters 5 to 7 deal with selection. Chapter 8 looks at breeding systems and the remainder of the book (Chapters 9 to 11) looks at practical situations and the future.

A deliberate attempt has been made to retain much of what Dr Dalton's book aimed to do, but additional areas are covered and some previously untouched at all are now introduced. An attempt has been made to keep mathematics to a minimum. It is hoped that this book will prove useful to agricultural students and veterinarians taking introductory courses, since they tend to be at the front line of large and small animal breeding respectively. It is also hoped that practical breeders will find the book of value and that all readers will be stimulated to take their knowledge further, to which end a bibliography of useful books is appended.

Malcolm B. Willis

Acknowledgement

I wish to thank my colleague Dr Mike Ellis who read the draft and made helpful suggestions. Any deficiencies that remain are mine and mine alone.

MBW

1 Traits in Farm Animals

Classes of traits

Trait, character or variable are terms applied to the features in which man has an interest with respect to his animals. Trait is the way we tend to refer to a particular feature, while character or variable are often how it is referred to in statistical terminology. Traits need to be very carefully defined or described, but essentially they fit into five basic categories.

(1) *Fitness traits*. These are usually concerned with reproduction and viability. Litter size, conception rate, calving interval, gestation length, survival ability are all examples.
(2) *Production traits*. These include such as milk yield, growth rate, feed efficiency, weaning weight. They need to be defined as to what part of the time scale is involved.
(3) *Quality traits*. Carcass composition, backfat thickness in the pig, eye muscle area in all carcasses, meat quality, milkfat percentage. These characters are increasingly important as pressures mount for more acceptable food in health terms.
(4) *Type traits*. These are usually features of a more aesthetic nature where personal preference is important, and the breeder selects for such things as coat colour, coat type, udder shape, physical appearance, horn shape, etc. Also included would be aspects of structural soundness such as legs/feet as well as such things as teat number in pigs.
(5) *Behavioural traits*. These are increasingly important in so far as animal welfare is concerned. They include not only the obvious ones such as herding ability in sheepdogs, but also docility in most species as well as response to stress. The ability of a sheepdog to herd sheep depends, for example, not only on the dog but also, in part, upon the instinct of sheep to flock together.

A basic principle of animal breeding is that the more traits included in any breeding programme, the harder the task becomes and the slower will be any improvement. It is imperative to decide on an order of merit or a list or priorities to adhere to, at least for some period of time. Priorities will, of course, change. In the early days of animal breeding purebred livestock breeders tended to place great emphasis upon type traits. Once scientists became involved they argued for production and quality traits, but the stress upon these latter has been more recently emphasised, while behavioural traits are more recent still. As pedigree breeds have become less important in certain species, type traits have become less vital. However, they are still considered crucial in some species and have economic merit to some breeders.

Most traits of importance tend to be polygenically controlled; that is, they are controlled by many genes but some, especially features affecting such things as coat colour, tend to be relatively simple Mendelian traits with only one or two major genes involved for each feature.

The importance placed upon particular traits should depend upon their economic value, and to this end production traits are generally of greatest importance as well as being among the easier ones to evaluate. In all species, reproductive traits have to be important since without good reproduction the species will be in difficulties as regards its survival. Litter-bearing species have great emphasis placed upon litter size or survival to weaning.

Measurement of traits

Traits can be classified as being either objective or subjective. Most live animal traits can be measured, e.g. weight gain, milk yield, fleece weight, litter size. Some are subjectively assessed using a scoring system or a hedonic scale, and this would apply to type evaluations as well as to some carcass attributes.

Breeding is frequently about making comparisons between one animal and another, or more particularly the traits measured in these animals. It is thus important that measurements are made on valid terms. It is, for example, difficult and unwise to compare animals reared and managed in different herds or to compare animals of differing ages.

Adjustment or correction factors have usually to be applied before some data can be legitimately analysed and comparisons made. Comparison within herds and years as well as age groups, such as is undertaken with contemporary comparisons for milk yield, is one way of eliminating herd/year/season and age effects, but this is not always feasible.

Some of the factors which need to be considered before comparisons can be made include:

(1) *Herd/year/season of year/location within country* might all be lumped together as management effects. Often these can only be reduced by making comparisons within specific groupings. Thus if animals born in the same herd in the same year and the same month are considered as contemporaries, and assessment made within this group, these effects are largely eliminated.

(2) *Age of animal.* Weaning weights of suckled calves will vary because calves tend to be weaned at a specific point in time but are born over a period of several months. Comparing weights of animals at differing ages is invalid so corrections need to be made to a constant age.

(3) *Sex of the animal.* This may be more important in some species than others (e.g. cattle more than sheep) but much depends upon the point in time. Males tend to be bigger and to grow faster and perhaps be leaner than females, so corrections are needed or the sexes should be assessed separately.

(4) *Litter size.* Individuals in large litters may grow less quickly than those in smaller litters, so adjustment to a constant litter size is called for or the use of covariance, a statistical technique for analysis which takes out the effects of litter size variation, is needed.

(5) *Dam age.* Young dams may produce slower growing progeny than older dams because of lower milk yields, hence the need to record and correct for dam age.

(6) *Parity.* This refers to the order of birth, e.g. first litter, second litter, etc., and although related to dam age is not necessarily identical. Parity must therefore be considered in analyses. Heifers, for example, are more likely to have difficult calvings than older cows.

Correction factors usually have to be calculated from the data being assessed though standardised factors are sometimes used. The validity of standardised factors is open to debate but need not concern us here. In general terms, correction factors are either additive or multiplicative.

In correcting the weaning weight of females to that of males, it is important to know not only the mean weights for each sex but the standard deviations. If males are heavier than females, but the standard deviations for weaning weight are more or less identical for the two sexes, then additive factors can be used. A fixed amount is added to each female weight to make it equivalent to what it might have been had it been a male rather than a female. If, however, females had not only a lower mean but a smaller standard deviation, then they should not be corrected additively but by a multiplying factor which would not only elevate the mean to the equivalent of males, but also enlarge the variation.

Traits of importance

Traits vary from species to species, they also vary in importance at different periods of time and to different segments of the agricultural industry. If the milk producer derives income largely from yield then total milk production will be more important than milk composition. If payment depends in part upon milkfat or protein content, then these assume greater importance to the animal breeder. The suckler calf producer selling calves at weaning will have less interest in post-weaning gain than the fattener buying those selfsame calves. If the national sheep industry is essentially one based upon meat production (as in Britain), then wool represents a relatively small part of the breeder's income. This contrasts sharply with Australia where wool is a principal source of income and selection has been directed towards that product rather than towards meat. A pig breeder is interested in litter size and litters produced per year but the buyer of weaned pigs has no interest in this, being more concerned with post-weaning gain.

Tables 1.1 to 1.4 show some of the economically important traits in different species. Not all will be equally important to all countries or to all areas of the industry, but all are of some importance to some. The heritabilities (see Chapter 5) are given only in broad terms because this feature is largely something peculiar to the data from which it was derived and thus detailed values are of less

Table 1.1 **Economically important traits of dairy cattle**

Trait	Heritability
Reproductive	
age at first calving	low
services per conception	very low
service period	very low
calving interval	low
twinning	low
dystocia	low
Yield per 305 day lactation (or per year or lifetime)	
milk	moderate
milkfat	moderate
protein	moderate
solids-not-fat	moderate
Composition (percentage basis)	
milkfat	high
protein	high
solids-not-fat	high
Other features	
somatic cell count (as indicator of mastitis)	moderate
aspects of type	range from low to high (mostly moderate)
rate of milk let down	low to moderate
temperament	low

importance but appear later (Table 5.1). In general, low means less than 15%, medium runs from about 20 to 50% and high refers to those above 50% though few high values would exceed 70%. High heritabilities indicate characters that are capable of showing faster progress, while low indicates characters that do not respond well to selection.

The lists presented in Tables 1.1 to 1.4 while comprehensive are not intended to be either exhaustive or in any order of priority. Indeed, the priorities may vary from breeder to breeder as well as from species to species. Priorities will depend in part upon economic considerations. Many scientists might argue that type characteristics

Table 1.2 **Economically important traits of beef cattle**

Trait	Heritability
Reproduction	
age at puberty/first mating	low
number of live births	low
dystocia	low
services per conception	low
Maternal ability	
weight calf weaned/cow	moderate
mothering ability	low
temperament	low
Weight	
birth	moderate
weaning (approx 200 days)	moderate
yearling	moderate
pre-weaning gain	moderate
post-weaning gain (feedlot)	high
400-day weight	high
Carcass traits	
killing out (dressing) per cent	moderate
fat depth (various locations)	moderate
longissimus dorsi area	high
classification/grade	moderate
weight of saleable meat	moderate
weight of excess fat	moderate
weight of bone	moderate
proportions of lean/fat/bone	low to moderate
ratio meat: bone	low to moderate
ratio first to second quality meat	low to moderate
muscling/shape	moderate to high
eating quality (tendernesss, etc.)	low to moderate
Other traits	
type aspects (various)	low to high (mostly moderate)
disease resistance	low
heat resistance (in tropics)	low to moderate

Table 1.3 **Economically important traits of pigs**

Trait	Heritability
Reproduction	
litter size	low
litters per year	low
number piglets born	low
survival ability (of piglets)	low
Weight	
weaning	low to moderate
slaughter	moderate
pre-weaning gain	low to moderate
gain weaning to slaughter	moderate
age at slaughter	moderate
Carcass	
carcass lean	moderate
killing out (dressing) percentage	moderate
fat depth (various)	moderate
longissimus dorsi area	high
carcass length	moderate
quality traits	low to moderate
halothane	single gene
Others	
type (rarely important)	low to high
teat number	moderate
temperament	low
skin colour	simple inheritance

should have a low priority, but many pedigree breeders would argue that animals of good physical appearance as regards their relationship to the breed ideal can command premium prices and thus type has economic merit for them. Others might argue that type characteristics are, in some instances, related positively to certain production traits, although this is more readily checked by scientific study.

Priorities will also differ according to the country in which the activity is taking place since economic weightings will vary by location. For example a hill sheep farmer in Britain will have fewer

Table 1.4 **Economically important traits of sheep**

Trait	Heritability
Reproduction	
number born	low
age at first mating	low
litters per year	low
Maternal ability	
milk production	low to moderate
temperament	low
survival ability of lamb	low
Weight	
birth	low to moderate
weaning	low to moderate
slaughter	moderate
pre-weaning gain	low
post-weaning gain	moderate to high
age at slaughter	moderate
Fleece	
weight	moderate
fibre diameter	moderate
staple length	moderate
follicle parameters	
Primary/secondary ratio	moderate
medullation (hairyness)	moderate
colour	moderate
Carcass	
killing out (dressing) per cent	moderate
classification/grade	moderate
fat depth	moderate
proportions (meat/fat/bone)	low to moderate
longissimus dorsi area	high
Others	
type and breed characteristics	low to high
	(usually moderate)

lambs per ewe than a lowland sheep farmer. Wool will thus be a greater proportion of the hill farmer's income as well as being more crucial to the survival of the lambs born in a harsher environment than those in the lowlands. Accordingly emphasis upon fleece type/weight will differ from the hill to lowland situation.

2 Mendelian Principles and Laws

The cell

All individuals of whatever animal species are made up of millions of minute cells. A typical cell is illustrated in Fig. 2.1.

Each cell contains a nucleus in which are found the chromosomes. These thread-like structures are constant in number within a given species though anomalies can sometimes occur as regards this number. Chromosomes come in pairs (called homologous pairs) with one member of each pair coming from each parent.

The numbers of chromosomes found in particular species are shown in Table 2.1.

The term *diploid* refers to the total number of chromosomes found in the animal's cells while *haploid* refers to the half number or number of pairs. The haploid number is found in *germ cells* (eggs and sperm).

Body cells have the ability to duplicate themselves by what is termed somatic growth which involves a process called *mitosis*. The body cells of, say, a cow will contain 30 pairs or 60 chromosomes. During mitosis the chromosomes split down their length into *chromatids*. These halved chromosomes are then drawn to opposite sides of the cell which then constricts between the separated chromosomes, and two cells result where one existed before. Each cell contains the full complement of 60 identical chromosomes.

In the formation of germ cells the process is called *meiosis*. Here there are two divisions, during the second of which only one member of each chromosome pair is drawn to each end. When the cell splits into two, each 'new cell' contains only one member of each pair of homologous chromosomes. Thus, unlike normal growth where the result is the production of duplicated cells, the germ cells are produced by a reduction division reducing the diploid chromosome

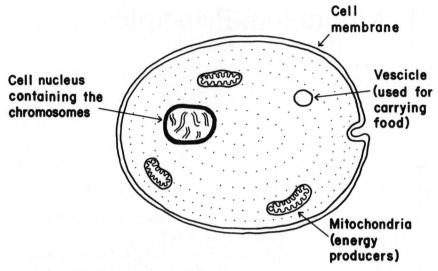

Fig. 2.1   **The main parts of a simple animal cell**

Table 2.1   **Chromosomes in various species**

	Chromosome number	
Species	*Diploid*	*Haploid*
Man	46	23
Pig	38	19
Rabbit	44	22
Sheep	54	27
Goat	60	30
Cattle	60	30
Bison	60	30
Donkey	62	31
Horse	64	32
Dog	78	39

number to the haploid. So when a bull's sperm bearing 30 chromosomes meets with the cow's egg containing 30 chromosomes, the resultant *zygote* once again contains the 60 chromosomes of cattle in the 30 pairs.

Simplified illustrations of mitosis and meiosis are shown in Fig. 2.2.

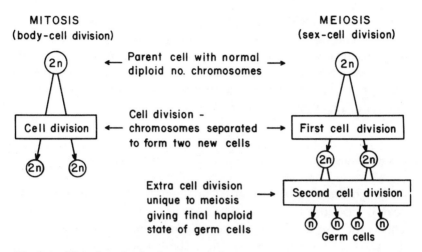

Fig. 2.2   **Meiosis and mitosis contrasted**

The method by which chromosomes end up in germ cells or *gametes* involves an element of chance. To be viable each germ cell should contain one member of each chromosome pair but whether it is the one originally from that animal's father or that originally from its mother is purely a chance occurrence. The chance element controlling which chromosome goes to which germ cell and the chance element of which sperm fertilises which egg leads to the chance of which genes an individual inherits.

Genes used to be likened to beads on a string with genes being the beads and chromosomes the string. We now know that genes are integral parts of the chromosome. Deoxyribonucleic acid (DNA) is the basic substance of most organisms. The DNA molecule was first proposed by Watson and Crick in 1953. DNA is composed of nitrogen-containing bases, purines and pyrimidines, a five carbon sugar (deoxyribose) and phosphate. The purines are adenine and guanine and the pyrimidines are thymine and cytosine. The difference between one gene version and another version of that gene (called an *allele*) is probably based upon the order of purines and pyrimidines.

Dominant alleles are those which 'work' to express themselves even when present on only one member of a set of homologous chromosomes. In contrast, a recessive allele needs to be present on both of the chromosomes in order to function. The place where a gene occurs is called a *locus* (plural *loci*) and a gene is always found at the same spot of the same chromosome pair. Chromosome mapping involves identification of the exact location of a specific gene on a specific chromosome. This is highly developed in *drosophila* (fruit fly) and is undertaken in a few other species but is not yet of great importance in domestic livestock.

Sex chromosomes

Although the concept of paired or homologous chromosomes has been discussed, one pair of chromosomes is not necessarily paired. This is the pair known as the sex chromosomes. In female animals the sex chromosomes are identical (both fairly large) whereas in males there is a large one akin to those in females, plus a small one. The female pair are termed X chromosomes (females are thus XX) while the male pair represent an X and a Y. In birds the sexes are the other way about and to avoid confusion, females are usually referred to as ZW, while males are ZZ. In fish the situation may be different again without specific sex chromosomes being identifiable.

Following the meiosis process, female eggs carry only one of these sex chromosomes and are thus all X. In contrast, male sperm will be either X or Y bearing. If a Y-bearing sperm fertilises the egg then the

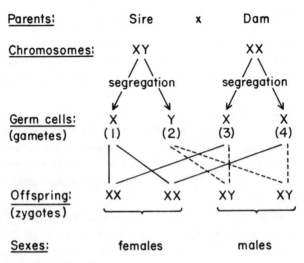

Fig. 2.3 **Determination of sex**

resultant offspring is male, and if it is an X-bearing sperm which makes contact first the resultant progeny will be female. This is shown diagrammatically in Fig. 2.3.

Although Y-bearing and X-bearing sperm are produced in more or less equal amounts, there is either some prediliction of the egg for Y-bearing, or Y-bearing are more motile. In any event, rather more XY zygotes are produced than XX, and although there is more mortality among male embryos a slightly increased number of males is seen at birth. Individual sires may produce abnormally high or low proportions of sons but in general terms, in many mammalian species, there are probably about 52 males in every 100 births. Sex ratio is the term used to define the number of males per 100 females; thus 106 would indicate 106 males per 100 females.

Sex chromosomes carry few genes of importance but such things as haemophilia A and B are produced by X-borne genes. Chromosomes other than the sex chromosomes are known as *autosomes*. Autosomal genes can be carried by both sexes whereas sex-chromosome X-borne genes may transmit from fathers to daughters and mothers to sons.

Although extra chromosomes occur in some individuals, this status is usually lethal. Exceptions would be the extra chromosome 21 in man which leads to Down's Syndrome. In many species additional sex chromosomes are also seen at fairly rare intervals. Thus one can have individuals which are XXY, XYY and XXX for example. These anomalies may exist in some body cells but not all, and usually there are related abnormalities generally with the sexual organs. It is known that in XX cases only one X chromosome is actually 'switched on' in any cell, though not the same one in all cells. This may account for the non-lethality of XXY or XXX individuals.

Mendel's law of segregation

Terminology used today may not be that used by Mendel but he did produce a rule or law which has stood the test of time. This was that the gene was the unit of inheritance and that genes exist in pairs in the cells of individuals. Germ cells carry one or other of those paired genes in equal proportions. Segregation involving one pair of genes (often called *monohybrid cross*) shows that genes remain discrete in

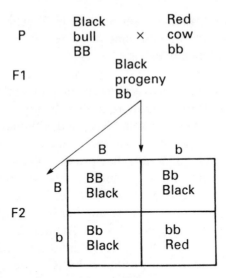

Fig. 2.4 **Segregation of gene with two alleles**

each generation and allow characters to reappear unchanged even if masked in one generation.

An example may be black or red colour in cattle. Black is dominant to red and thus we use the upper-case letter B to signify black and the lower-case b to signify red. Since only two alternatives exist we have only two phenotypes: Black or Red. However blacks can be either BB or Bb while reds are always bb. We thus have three genotypes. If we cross a pure-breeding black bull (BB) to a red cow (bb) then all the progeny would be black (Bb). However if we mate together individuals from this first cross or F1 generation we obtain a fixed proportion of blacks and reds in the next (F2 generation). This is shown diagrammatically in Fig. 2.4.

In the F2 generation we obtain three blacks to every red or a 3:1 ratio. However, closer examination shows that only one of the blacks is pure-breeding (BB) while two are carrying the red factor (Bb).

Had we used an example with Shorthorns using Red or White genes, with R representing red and r representing white, then the pattern would be identical to that in Fig. 2.4 substituting R for B and r for b. However in this case, the Rr individual is not red but a roan colour; a mixture of red and white hairs. This is not blending of

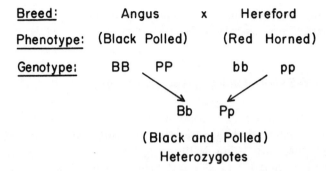

When these heterozygous animals are crossed the results are these:

Male

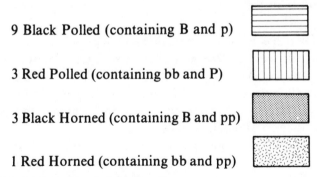

The phenotype ratio from this is:

9 Black Polled (containing B and p)

3 Red Polled (containing bb and P)

3 Black Horned (containing B and pp)

1 Red Horned (containing bb and pp)

Fig. 2.5 **Segregation of two independent genes**

genes but indicates codominance or no dominance since R does not mask r in the way that B masked b.

Mendel's law of independent assortment

Mendel's so-called second law is concerned with the *dihybrid cross*. This states that genes affecting different traits will segregate independently. This is illustrated in Fig. 2.5 for matings between black polled cattle and red horned cattle. In this case black and polled are dominant to red and horned respectively.

Of the 16 types seen in the F2 generation nine would show both dominant features (black/polled), three would show one dominant feature (black/horned) and three the other dominant feature (red/polled) while one would show both recessive features (red/horned). We thus have a 9:3:3:1 ratio.

With three genes each with two alleles there would be a 27:9:9:9:3:3:3:1 segregation. All three dominant features would be shown by 27 animals; there would be nine showing each of two features, three showing each of one dominant feature and one showing all three recessive features.

Multiple alleles

So far it has been assumed that only two alleles exist at any gene locus. In reality some genes have several alternatives and are thus multi-allelic. An individual can only carry two versions; either the same (*homozygous*) or different (*heterozygous*) but in the population at large all versions may exist and segregation will still work according to Mendelian rules.

The ABO blood group in man is a typical example. Individuals can be AA, AO, BB, BO, AB and OO in genotype. Since A and B are dominant to O the AA and AO genotypes appear to be A phenotypically and the BB and BO appear to be B phenotypically. There is no dominance of A over B so AB is quite distinguishable as is OO. Individuals cannot carry A, B and O because there are only two chromosomes and thus only two alleles are feasible. Similar genes exist in animals especially with regard to coat colour. Sometimes they are marked with suffixes, e.g. a^t or a^s or A^y.

Because gene numbers can be very large a general formula is needed. If there are n genes then there can be 2^n gametes and 3^n genotypes. Even with only 20 genes each with two alleles the number of genotypes runs into over a thousand million. Aside from identical twins the genotype of an individual is thus something totally unique to that individual and no two animals are alike in genetic make-up.

Lethal genes

Some genes act in such a way that they cause the individual carrying them to die either before or subsequent to birth. In the case of a defect that is lethal in embryo, there is often an abnormal segregation ratio. Frequently the lethal is a recessive trait such that the aa type is non-viable and a 1:2 ratio of AA to Aa is seen with no aa. At times the dominant gene may be lethal when in duplicate so that the AA animal does not appear and the ratio is 2:1 for Aa:aa. Von Willebrand's disease (a blood disorder) is believed to be lethal in the double dominant state as is dominant white coat colour in some types of horse. Usually, however, defects that are dominant have been bred out of species and most lethals tend to be recessive.

Examples of lethal genes include the dropsy syndrome (bulldog) seen in Ayrshire cattle; amputate seen in Friesian cattle, A–46 (underdevelopment of the thymus) in Black Pied Danish cattle; imperforate anus seen in pigs and sheep; hydrocephalus in most species, and hairlessness in most species.

Defects that do not actually cause death but lead to serious imperfections are often called semi-lethals, but most badly deformed animals would be culled even if they did not actually succumb to the condition they have unfortunately inherited.

Linkage

An exception to the Mendelian law of independent segregation concerns linkage. This occurs when particular genes are at different loci but are carried on the same chromosome. Because chromosomes tend to be transmitted in their entirety, the different genes are transmitted together. Segregation rules may thus not hold good.

Chromosomes do, however, tend to crossover with others, and

parts become interchanged so that a chromosome may be transmitted minus, say, the last quarter but instead transmit the last quarter of its homologous partner. The chance of 'breaks' are obviously greater the longer the segment of chromosome being considered. Different genes situated close together on a particular chromosome tend to be linked more than those situated at opposite ends of the chromosome. Linkage situations are very valuable in detailing chromosome mapping. Although linkage may inhibit new combinations it enables existing ones to be retained.

Sex-linked, sex-limited and sex-influenced traits

As previously mentioned sex-linkage refers to characters linked to the sex of the individual because they are carried on the sex chromosomes (usually the X). Haemophilia A is one such trait where we could signify H as the normal clotting factor while h is the recessive allele causing problems at Factor VIII of blood. Females can be HH, Hh or (rarely) hh; all of these on the X-chromosome. The last of these would be affected and the second one would be termed a carrier.

In contrast males having only one X chromosome are either H or h with nothing on the Y chromosome. Males that are h are affected. Accordingly they will, if used, pass the affected X(h) chromosome to *all* their daughters but a normal Y goes to their sons. These carrier daughters will, in turn produce some affected sons. To obtain affected females one would have to mate an affected male to a carrier female which is inevitably a rare event.

Sex-limited traits are not carried on the sex chromosomes but are autosomal. However they can only be expressed by one sex. Milk yield and egg production are obvious sex-limited traits seen only in females. In contrast, cryptorchidism is a male-expressed trait that cannot be seen in females. Although seen only in one sex, these sex-limited traits are transmitted by both, hence the importance attached to the progeny testing of dairy bulls or cockerels.

Sex-influenced traits are again autosomal but occur in both sexes. However there is a preponderance of cases in one sex rather than the other. Some examples of horns in sheep are sex-influenced in that identical genotypes may produce polling in females but some indication of horns in males. Mahogany red in Ayrshire coat colour is more likely in males than females. By the same token, hip dysplasia, a condition seen in the dog and to a lesser degree in Hereford cattle, is more frequent in females than males as it is in man.

Epistasis

Although different genes segregate independently (linkage excepted) it is possible for a gene at one locus to have an effect upon the expression of another gene at a totally different locus. This is termed epistasis and it differs from dominance in the sense that dominance refers to alleles of the same gene while epistasis refers to different genes.

An example of this is coat colour in Labrador retrievers which has three basic phenotypes Black, Yellow and Chocolate (Liver). Black is caused by the dominant allele B with the recessive version b giving rise to chocolate coloration. However, in order for these alleles to function there needs to be the dominant gene E from the extension series. The alternative to this is the recessive allele e which causes black pigment to fade from the coat and thus give rise to yellow. Although ee causes yellow pigment in the coat it does not influence nose colour but bb turns black pigment in the coat and nose to liver. With the possibilities of B or b coupled with E or e there are nine basic genotypes as shown in Table 2.2.

Table 2.2   **Epistatic effect on coat colour in Labradors**

Phenotype	Genotypes			
Black	BBEE	BbEE	BBEe	BbEe
Chocolate	bbEE	bbEe		
Yellow (black nose)	BBee	Bbee		
Yellow (liver nose)	bbee			

3 Principles of Population Genetics

Economic traits

In Chapter 1 traits of possible economic importance in farm livestock were listed for each of several species. Most of these traits are not only complex in terms of the factors which influence them, but are also complex in their mode of inheritance. Unlike most traits mentioned in Chapter 2 the important traits of farm livestock tend to be controlled by many as opposed to single genes. Such traits are termed polygenic in being influenced by a large number of genes, each of which is individually of minor importance but which collectively have a cumulative importance. In addition, environmental factors will play their part. This is illustrated in Fig. 3.1 which shows the factors that might influence a trait such as 400-day weight in beef cattle.

Variation

Anyone dealing with biological data realises that there is variation in whatever characteristic one examines. Some traits show minimal variation, e.g. calf numbers born to beef cows will range over the area 0, 1 or 2 and only rarely exceed the level of twins though 3, 4 and 5 calves are known to occur. In contrast, the trait of 400-day weight will vary over a quite large range.

For purposes of illustration assume that 400-day weight is controlled by three pairs of genes each with two alleles A/a, B/b and C/c. Also assume that the basic combination aabbcc gives a 400-day weight of 350 kg and that each upper-case letter (whether A, B or C) adds 20 kg to this basic weight. Further assume that environmental influences are non-existent. An individual of genotype AABBCC

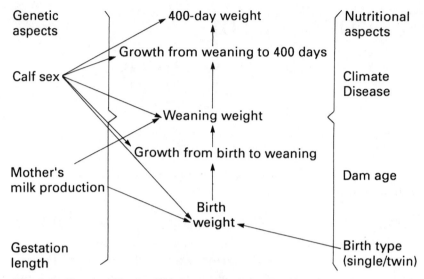

Fig. 3.1   **Factors affecting 400-day weight in beef cattle**

will then weigh 470 kg compared with an individual weighing 350 kg
for genotype aabbcc. Consider a mating of these two.

The resultant progeny will be of genotype AaBbCc and will thus
have a weight of 410 kg which is midway between the parental
weights of 470 kg and 350 kg. Now mate individuals of genotype
AaBbCc together. These represent the F1 generation and will give
rise to the F2 generation. The A/a, B/b and C/c alleles will segregate independently and the F2 generation will appear as shown in
Table 3.1.

The 64 individual types seen in the F2 generation exhibit all the
weights (in 20 kg units) from the original parental low of 350 kg to
the original parental high of 470 kg. The frequencies are shown in
Table 3.2. This illustrates considerably more variation in the F2
compared with the F1 but it will be noted that the overall mean is the
same in both the F1 and F2 generations. It will also be seen that
although there are 64 individuals they divide into 27 distinct
genotypes.

The 27 different genotypes in Table 3.2 actually show segregation
into seven different phenotypic categories. Expressed another way
there are 27 individuals showing all three upper-case letters (A, B
and C). There are nine individuals in each of the AB, AC and BC
categories. There individuals each exhibit either A, B or C and one

Table 3.1 **Segregation of three genes affecting 400-day weight (kg) in beef cattle (hypothetical)**

Male gametes	Female gametes							
	ABC	ABc	AbC	aBC	Abc	aBc	abC	abc
ABC	AAB BCC 470	AAB BCc 450	AAB bCC 450	AaB BCC 450	AAB bCc 430	AaB BCc 430	AaB bCC 430	AaB bCc 410
ABc	AAB BCc 450	AAB Bcc 430	AAB bCc 430	AaB BCc 430	AAB bcc 410	AaB Bcc 410	AaB bCc 410	AaB bcc 390
AbC	AAB bCC 450	AAB bCc 430	AAb bCC 430	AaB bCC 430	AAb bCc 410	AaB bCc 410	Aab bCC 410	Aab bCc 390
aBC	AaB BCC 450	AaB BCc 430	AaB bCC 430	aaB BCC 430	AaB bCc 410	aaB BCc 410	aaB bCC 410	aaB bCc 390
Abc	AAB bCc 430	AAB bcc 410	AAb bCc 410	AaB bCc 410	AAb bcc 390	AaB bcc 390	Aab bCc 390	Aab bcc 370
aBc	AaB BCc 430	AaB Bcc 410	AaB bCc 410	aaB BCc 410	AaB bcc 390	aaB Bcc 390	aaB bCc 390	aaB bcc 370
abC	AaB bCC 430	AaB bCc 410	Aab bCC 410	aaB bCC 410	Aab bCc 390	aaB bCc 390	aab bCC 390	aab bCc 370
abc	AaB bCc 410	AaB bcc 390	Aab bCc 390	aaB bCc 390	Aab bcc 370	aaB bcc 370	aab bCc 370	aab bcc 350

Table 3.2 **Gene combinations for 400-day weight (hypothetical)**

Gene combination	Number	400-day weight (kg)
AABBCC	1	470
AABBCc	2	450
AABbCC	2	450
AaBBCC	2	450
AABBcc	1	430
AAbbCC	1	430
aaBBCC	1	430
AABbCc	4	430
AaBBCc	4	430
AaBbCC	4	430
AABbcc	2	410
AAbbCc	2	410
AaBBcc	2	410
AaBbCc	8	410
AabbCC	2	410
aaBbCC	2	410
aaBBCc	2	410
AabbCc	4	390
AaBbcc	4	390
aaBbCc	4	390
aabbCC	1	390
aaBBcc	1	390
AAbbcc	1	390
aabbCc	2	370
aaBbcc	2	370
Aabbcc	2	370
aabbcc	1	350

individual shows only the aabbcc combination. This is the exact segregation described in Chapter 2 in the section *Mendel's law of independent assortment*. However, in the 400-day weight example the alleles A, B and C were considered to be identical in their action. If the letters are looked at as upper- or lower-case regardless of the actual letter than Table 3.2 can be reassessed in histogram form looking simply at how many upper-case letters are present. This is shown in Fig. 3.2.

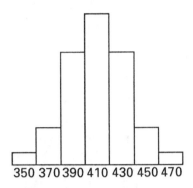

350 370 390 410 430 450 470

Fig. 3.2 **Histogram of 400-day weights in Table 3.2**

Although the divisions in Fig. 3.2 are quite broad (20 kg) the pattern exhibited is one in which there are few animals at the extremes and more towards the centre. This is a typical pattern seen in polygenic traits and if the top points of the histogram are joined in a continuous line the pattern approximates to what is called a normal curve.

The 400-day weight example was an oversimplification in that many more than three gene pairs are involved in the genetic determination of this character. Moreover, no account was taken of sex differences nor of any environmental influences. The fact that environmental influences act upon the genetic make-up to modify the phenotype means that the smooth balanced pattern of the histogram is modified. There would, in reality, be much smaller divisions than 20 kg and a much larger spread of weights. The example illustrates something about variation which is a fundamental feature of polygenically controlled traits.

Animal breeders are interested in seeking uniformity and there are many practical advantages if individuals all perform in similar fashion. The nature of genetics plus environmental influences make this impossible and a geneticist would regard variation in both the visible (phenotype) and invisible (genotype) sense to be highly desirable if selection is to be successful. In simple terms, population (quantitative) genetics may be said to be primarily about variation.

The easiest way to illustrate variation is to show the distribution of a particular trait in diagrammatic form. Figure 3.3 shows the birth weights of single-born Scottish Blackface sheep out of two-year-old mothers. Fairly broad divisions have been used and the balance each

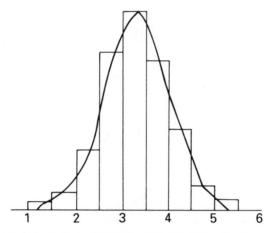

Fig. 3.3   **Birth weights (kg) of 1290 Scottish Blackface lambs**

side of the centre which was seen in the hypothetical 400-day weight case is partly lost. Nevertheless these real data show the same normal distribution pattern expected of most important traits.

Features of normal curves

Normal curves, regardless of the trait for which they are drawn or the units of measurement, show certain features which always apply if the curve is truly normal. The curve is bell shaped with the highest point around the mean and decreasing numbers as the extremes are reached. Curves can vary considerably. They can be very spread out along a large base or narrow with a high peak and a small base.

All normal curves have a variation which in statistical terms is called variance (σ^2). If we call each individual X and the number of individuals n then the variation of a population is given by the formula:

$$\sigma^2 = \frac{\Sigma X^2 - \Sigma(X)^2/n}{n - 1}$$

A high variance would indicate a very variable character and a low variance a higher peaked type of curve with a small spread along the baseline. However variance is expressed in terms of the items squared. Variance measures the deviations of each item from the overall mean but kilograms squared is not a very meaningful term. Of more meaning than the variance is the standard deviation. This is the square root of the variance and thus is expressed in the actual units of measurement. If 400-day weight in bulls of a particular breed had a mean of 500 kg ± 30 kg where 30 kg was the standard deviation then it would signify something about the nature of the population concerned.

In any normal curve the standard deviation describes the variability. It is established that one standard deviation either side of the mean encompasses about 68% of the population while two standard deviations will enclose about 95% and three will encompass about 99%. This applies to any normal curve regardless of trait or species being examined.

In the bull example stated a mean of 500 ± 30 would indicate that about 68% of bulls weighed between 470 kg and 530 kg, while 95% would weigh between 440 kg and 560 kg, and 99% between 410 kg and 590 kg. A bull weighing 680 kg would be six standard deviations above the mean and this would be such a rarity that it would almost certainly signify that this bull was of a different breed to the main population.

Normal curves of the type discussed are seen for most important traits and these characters are often called *continuous variables*. Most characters that are polygenic in their mode of inheritance follow these patterns but not all. Fertility, when measured in

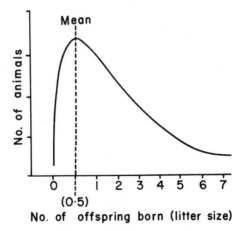

Fig. 3.4   **A skewed distribution (litter size in sheep)**

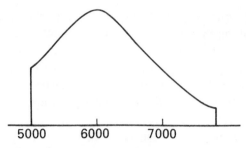

Fig. 3.5   **Truncated curve of milk yield (litres)**

numbers of offspring born, may follow this kind of pattern for litter producers like pigs or dogs, but that would not apply to cattle or sheep. Figure 3.4 shows a more typical pattern for sheep – it is skewed in that it has a long tail towards the upper end. Normal curve statistics would not apply to such populations which generally require the data to be transformed into some other format such as logarithms before analyses can be reliably undertaken.

Traits like litter size in sheep are called *discrete variables* and different statistics apply. The same is true for selected populations. Figure 3.5 illustrates a population from a herd book in which entry to the book requires the animal to have given a specific yield of milk. The curve is a kind of truncated normal curve. Yields less than 5000 litres will exist but such cows are excluded. Application of normal curve statistics to such populations would be incorrect. It is a prerequisite for studies of genetic parameters that the population being used is a normal one and not one that is biased in any way.

Animal breeders are working in the main with populations which display normal variation, and in which the objective is to bring about improvement. There may be considerable debate about which characters to improve and why, but the objective is essentially always the same – namely to improve the mean value of that trait(s). In essence the animal breeder is seeking to change gene frequencies in the population by increasing the frequency of desirable genes and reducing the frequency of the undersirable ones. The problem is that, apart from simple Mendelian traits, breeders do not know how many genes are involved, still less are they able to identify them. They are therefore working in a kind of empirical fashion. This does not mean that animal breeding is a hit and miss affair, although there is clearly some element of chance (luck) involved. Although genes cannot be identified, traits can be measured or evaluated and certain standard techniques can be used to bring about progress.

Many advances in the development of breeds and types of livestock were undertaken long before Mendel, and it is often claimed as a consequence that 'practical' breeders were better than 'scientists'. Practical breeders tended to work by certain rules of thumb for which genetic support is now sometimes forthcoming. For example the old adage 'breed the best to the best' meant breed the best males to the best females, with 'best' referring to excellence in some specific feature, usually appearance. This is fairly sound genetic practice for polygenic traits although it will not always apply, and with simple Mendelian traits it will be realised that the 'best to the best' will not always work for a variety of reasons.

Threshold characters

Although most polygenic characters show a normal curve type of distribution there are characters which exhibit an 'all or nothing' type of appearance akin to a Mendelian trait but which are controlled by many genes. Survival of the young is one such trait where there are only two versions (survival or death) but where the underlying genetic pattern may be complex. Cryptorchidism is another such feature with the added complication that it is seen only in males. Twinning could be considered as being an all or nothing trait, in that a cow either does or does not give birth to twins.

Although such traits may exhibit discrete patterns as opposed to the pattern of a normal curve it is known that some of these characters are actually controlled by many genes. The principle seems to be that there will be a specific number of genes at which the phenotypic expression changes. Suppose, for example, that cryptorchidism had five genes influencing it, each with two alternatives (0 or +), the point at which change ensued might be at the stage of seven-plus alleles. All individuals with fewer than seven-plus alleles would exhibit the normal state, but those with seven or more plus alleles would exhibit the cryptorchid state. The point at which the phenotypic change will occur is termed the threshold point, hence these characters are called threshold traits.

Some characters have several divisions, thus patent ductus arteriosus (a heart disease) occurs when the ductus arteriosus (a foetal

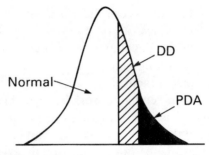

Fig. 3.6   **A threshold pattern**

condition) persists after birth. Normally this duct closes post-natally but if it only partially closes it will be termed ductus diverticulum, whereas if it remains open it is patent ductus arteriosus or PDA. Figure 3.6 illustrates this.

Threshold traits are particularly difficult both as a concept and in terms of seeking to influence them. If only normal and abnormal versions exist then both are recognisable and culling of the abnormal may be easy. However one cannot readily distinguish between those normals which are very çlose to the threshold point and hence potentially dangerous, and those normals which are far removed from the threshold point and thus unlikely to give rise to abnormal offspring.

Sometimes all or nothing traits can be classified into a series of categories as opposed to merely two or three, and this aids in the analysis and management of the problems. Hip dysplasia, a defect seen in Border Collies as well as other breeds of dog, (also in cattle, horses, rabbits, cats and man) can be classified as normal, near normal or affected. It has also been assessed on a scoring system which changes it from a trait with three discrete sections into something approaching a normal population with animals scoring from 0 through to 106 on the basis of radiographic features of the hip (see Willis, 1989).

Ways of selecting threshold traits and assessing genetic parameters are given in Chapter 5.

4 Ways of Changing Gene Frequencies

Introduction

Animal breeders are interested in 'improving' their animals but there may be considerable disagreement about what constitutes improvement. Some breeders will seek to alter production traits, others will be looking for improved physical appearance, yet others for combinations of these. The features they are examining are complex characters which are controlled partly by the animals' genes and partly by environmental features. What breeders are actually doing is altering the frequencies of particular genes even if they may not know which genes they are dealing with. A breeder seeking to improve weaning weight of beef cattle may be weighing cattle at weaning and selecting the heaviest. In part selection will be for animals which produce more growth hormone per unit of body weight and thus will increase the frequency of genes controlling this feature and reduce the frequency of other genes. To understand the principles it is useful to look at the single gene situation which involves awareness of the Hardy–Weinberg law.

The Hardy–Weinberg law

This law was first discovered in 1908 and bears the names of the two men who independently documented it. Essentially it states that in a large population the gene frequencies will remain unchanged if there is:

(1) Random mating (each animal has equal opportunity of mating any animal of the opposite sex).
(2) No mutation.

(3) No migration.
(4) No selection.

If the gene frequencies do not change then the population is said to be in equilibrium. The essential features of this are a large population, random mating and no 'changing forces'.

It is not within the scope of this book to go into mathematical proofs of the Hardy–Weinberg law or others but understanding of the principles involved has a bearing upon the understanding of how to change populations in the desired direction.

Assume a randomly mating large population of Friesian cattle with no forces seeking to change gene frequencies and in which the red coat colour, as opposed to black, appears in 1% of calves born. For purposes of this example the white markings also seen with the black or red coat colour can be ignored.

The Hardy–Weinberg law calculations suggest that in a situation with two alleles (B or b) and therefore three genotypes (BB, Bb and bb) the frequencies of the genotypes will be:

BB	Bb	bb
p^2	$2pq$	q^2

where q is the frequency of the red allele (b) and p that of the black (B) and $p + q = 1.0$.

If there are 1% red calves then bb or q^2 is expressed as 0.01. If q^2 is 0.01 then q, its square root, is 0.10. Since $p + q = 1.0$ it follows that $p = 1.0 - 0.10 = 0.90$.

The frequencies of the three genotypes would then be:

BB (0.81)   Bb (0.18)   bb (0.01)

Because we are dealing with a dominant gene causing black coat colour, we cannot distinguish between the BB and Bb animals in appearance since all are black. We can only say that 99% of the cattle are black and 1% are red. However the Hardy–Weinberg calculations suggest that 18 of the 99% blacks actually carry the red gene recessively. In making these calculations it is assumed that there is no selection. In reality red coat colour may be selected against to some degree by pedigree breeders so the calculations we have made will overestimate the number of Bb cases. Nevertheless as a rough guide Hardy—Weinberg is quite effective.

The example used here relates to a simple arrangement involving an autosomal recessive trait but the Hardy—Weinberg law applies equally to sex linked and/or multiple allele loci. Any theoretical genetic text book will deal with these cases.

Assessing deleterious genes

The Hardy—Weinberg law is often used when one is seeking to assess the frequency of undesirable alleles in a population. Suppose that dropsical calves occurred in an Ayrshire cattle population at the rate of one every 300 calvings then what would be the incidence of carriers in the population? An appearance of 1 in 300 is a frequency of 0.0033. It is established that dropsy is a simple autosomal recessive so that 0.0033 represents q^2. This means that q = 0.0577 and therefore p is $1 - 0.0577 = 0.9423$. The proportion of carrier animals would be 2pq or $2 \times 0.9423 \times 0.0577 = 0.1088$. There would thus be a situation in which about one animal in every ten is a carrier.

Factors changing Hardy—Weinberg frequencies

MUTATION

A sudden heritable change in genetic material is called a mutation. Any change that occurs once is likely to be lost unless the resultant genotype has a selectional advantage, but the same is not true of a recurrent mutation which is going to occur at a regular frequency each generation.

Mutation rates tend to be very low ranging from around 10^{-4} to 10^{-8} for most cases. The effect of such rare events will tend to be of minimal significance but over a long time could have a slight effect. If allele A mutates to a at the rate u and allele a mutates to A at the rate v then eventually when pu = qv there will be a point of equilibrium. If the mutation is in one direction only then the mutating allele will be gradually declining.

Because mutation rates are usually so low mutational effects upon the Hardy–Weinberg frequencies are rarely of major importance.

MIGRATION

Of far greater importance than mutation is migration which is effectively the movement of individuals from one population to another. In animal breeding terms the practice of grading-up (when sires of a second breed are used in successive generations in an existing breed) is an example of migration. The virtual demise of the Shorthorn in Britain from the mid-1950s did not occur by the removal of the breed but rather by the successive introduction of Friesian sires to 'grade-up' and, in the end, almost eliminate the Shorthorn. The same principle has occurred in different breeds, species and locations.

The effect of migration upon Hardy–Weinberg frequencies is given by the formula:

$$\Delta q = m (q_m - q_o)$$

In this equation m is the proportion of migrants and q_o and q_m are the gene frequencies of the original and migrant population respectively. The change of frequency is indicated by Δq.

In effect the migrational influence depends upon how many migrants are used and how different those migrants are in gene frequency from the original population.

Assume a cattle population that is red in coat colour and thus all bb into which black bulls (BB) are used at the rate of 15% of the matings. The original frequency of b in the population (q_o) was 1.0 and that in the migrants (q_m) is 0.0. The value for m is 0.15. Since black bulls replace 15% of the red ones the formula for change becomes:

$$q_1 = 0.15/2(0.0 - 1.0)$$
$$= -0.075$$

The division of 0.15 by 2 occurs because only males are introduced, not females. In one generation the red allele has thus declined from 100 to 92.5% through migration. Further use of black bulls in future generations would continue to reduce frequency of the red allele.

SELECTION

Whether artificial or natural, selection can have a marked effect upon frequency, although not necessarily as marked as some forms of migration. However migration is only effective up to a given point, whereas selection can be an on-going feature. When considering selection at the single gene level it is important to define the type of gene action and the selection being practised. Selection may be against a recessive allele or against a dominant allele, or could be in favour of the heterozygote, or there could be sex-linked aspects to consider. It is beyond the scope of this book to go into great detail on all of these possibilities which are effectively dealt with in theoretical books listed in the Further Reading section. It does however seem appropriate to look at selection against a recessive allele, because in many livestock species practical breeders are concerned with trying to eradicate deleterious genes (usually recessive).

In selecting against or for a gene one has to look at the phenotypes and their relative 'fitness'. The term fitness in a naturally selected population relates to the capacity of the particular phenotype to survive and reproduce. With farm livestock man tends to make the decisions about which animals will be retained for breeding, and this selection by man is termed artificial selection. However it does not always follow that these plans come to fruition. Animals selected as potential breeding stock may die or prove to be infertile which means that natural selection still has some influence upon breeding actions. The term 'fitness' is thus not quite the same as in natural selection but relates to those phenotypes which the breeder thinks to be the most suitable for reproduction in the context of what the breeder is trying to achieve.

In a situation of a dominant gene with selection against the recessive allele there are three genotypes: AA, Aa and aa. Because AA and Aa are phenotypically indistinguishable they are, in practice, given equal importance, whereas aa is selected against. The theory is laid out in Table 4.1.

The gametic contribution refers to the contribution which specific genotypes make to the next generation. It is calculated by multiplying the frequency by the fitness.

A fitness of 1 is given to the desired phenotype (though not all may survive to reproduce) and s (the coefficient of selection) determines the extent of the selection against the unwanted phenotype. If the aa

Table 4.1

Phenotype	AA	Aa	aa	Total
Frequency	p^2	$2pq$	q^2	1
Fitness	1	1	$1 - s$	
Gametic contribution	p^2	$2pq$	$(1 - s)q^2$	$1 - sq^2$

phenotype was lethal than s would equal 1 and all aa animals would be culled. An s of 0.30 would indicate that for every 100 animals of the favoured phenotype used then 70 of the unfavoured would be used (i.e. a selection of 30% against the unfavoured phenotype).

Although the original frequency totals 1, the gametic frequency total becomes $1 - sq^2$ because some of the aa individuals are culled from breeding. The frequency of the A allele in the next generation (p_1) becomes:

$$P_1 = \frac{P_o}{1 - sq_o^2}$$

and the frequency of the a allele becomes $1 - p_1$.

Consider the red allele in the Friesian and suppose it exists at the level of 1% red calves born, which is a frequency of b of 0.10, and supposed no red animals will be bred from: then $s = 1$.

This situation is as shown in Table 4.2.

The frequency of p(B) is 0.90 and that of q(b) is 0.10. The frequency of p in the next generation will be:

$$P = \frac{0.90}{1 - 1(0.10)^2} = 0.909$$

Table 4.2

Phenotype	Black	Black	Red
Genotype	BB	Bb	bb
Frequency	0.81	0.18	0.01
Fitness	1	1	0

and the frequency of q will be $1 - 0.909 = 0.091$ so that red calves will appear at the rate of 0.091^2 or 0.0083. In the original generation there were 1% red calves which has declined to 0.83%, a slight reduction.

Selection against a recessive of this nature is not very effective, not because selection against it is not severe (to cull all red calves is making it as severe as a lethal) but because the incidence of the red allele was low.

It is easier to select against alleles that have a high frequency than those that are low, and as selection against a rare allele continues it becomes harder to reduce it still further.

In many small animal species great store is laid upon 'culling defects' but the reduction of incidence of a rare defect is a laborious and not always very fruitful occupation.

It can be shown that to reduce a gene from q_o to q_t would take:

$$\frac{1}{q_t} - \frac{1}{q_o} \text{ generations}$$

Consider the dropsy gene with a frequency of 1 in 300 where it is hoped to reduce this incidence to half its current level (i.e. 1 in 600) by not using any dropsical animal (in the vast majority of cases they are not viable anyway). It can be shown that the original frequency is 0.0033 or a q of 0.0578 and the intended frequency is 0.0017 or a q of 0.0408. This would take a total of 7.2 generations to achieve. If generation interval (see Chapter 5) in Ayrshires is about six years then the task would cover the best part of 43 years. Despite all this effort dropsy would not have disappeared, simply declined in incidence. ,

Detecting carriers of deleterious recessives

As shown earlier most species carry deleterious genes. In man over 1500 simple genes exist which are deleterious in their mode of expression, and it is not improbable that similar numbers exist in other species, although less attention is given to such features in farm livestock than in man or small animals like the dog or cat. Nevertheless, animal breeders often wish to know if particular animals carry such defects which are generally, though not exclusively, simple autosomal recessives. Various techniques exist to determine carriers of defects and these are discussed in turn.

Known or likely carriers of recessive alleles can sometimes be predicted from pedigree data. If the animal being studied shows the dominant allele (e.g. N) then the 'worst' it can be is Nn but it may be NN. If one parent was nn then it is obvious that the animal must be Nn, since that particular parent can only have given n to its offspring. Similarly, if both parents had been known to previously produce nn progeny but were themselves N, then it follows that these parents must both be Nn and the animal in question has a 2:1 chance that it is Nn and a 1:2 chance that it is NN. Simple Mendelian ratios coupled with chi-square tests can be used to estimate ratios and chances from pedigree data. However, in many cases pedigree data may be totally absent and other techniques are required.

Test mating is usually regarded as the mating of a suspect 'carrier' (N?) to either a known carrier (Nn) or an affected individual (nn). The most reliable method is to use nn individuals for testing, but this is only feasible if the nn genotype is viable (as it would be with a coat colour defect but not with more serious recessives). If the N? male is mated to the nn female and an affected individual (nn) results then the sire is proven to be Nn. However failure to produce an nn case does not exonerate the N? individual and an estimate of risk needs to be made from various matings.

Test mating of this kind is only useful if the affected parent can be bred from and if the defect is fairly early in onset. There is little point in test mating a sire for a defect that does not appear in the offspring until they are middle-aged since by that time the sire being tested might well be either deceased or no longer fertile.

In cases where the homozygous female cannot be used an alternative is to use proven carrier females (Nn). Because the risk of affected is less in the N? × Nn mating than in the N? × nn mating more progeny are required by this technique to exonerate a sire to a given level of probability.

Two other methods of test mating would be to mate the suspect sire to daughters of a known proven carrier (Nn) or to mate the suspect sire to his own daughters. Both have equal probabilities in terms of identification of carriers and the differences between the two techniques are mainly practical ones. The former of these techniques is clearly quicker than the latter. However, mating a sire to his own daughters will not only test him for the problem under study but may throw up other recessive problems carried by him but hitherto unsuspected.

Table 4.3 shows the relative probabilities of being wrong given specific numbers of progeny from each of these four methods.

The figures given in Table 4.3 represent the probability of being wrong if one assumes NN status. The probability of detecting a sire

Table 4.3 **Test mating errors (%) when testing for a simple autosomal recessive. Percentage chances of being wrong when assuming N? to be NN because no nn progeny resulted**

Normal progeny seen	Mating to affected female (nn)	Mating to carrier female (Nn)	Mating to daughters of proven carrier or sire's own daughters
1	50.0	75.0	88
2	25.0	56.3	78
3	12.5	42.2	71
4	6.25	31.6	66
5	3.13	23.7	62
6	1.56	17.8	59
7	0.78	13.3	57
8	0.39	10.0	55
9	0.20	7.51	54
10	0.10	5.63	53
11	0.05	4.22	52
12	0.02	3.17	52
13	0.01	2.38	51
14		1.78	51
15		1.34	51
16		1.00	51

as a carrier is thus the reciprocal of these figures, e.g. with six progeny to an nn female there is a 98.44% chance of detection. It does not matter if progeny are individuals or in litters. When assessing relative risks by mating to nn or Nn individuals the chances assessed are additive. Thus a litter of eight individuals would be the same as eight separate matings, each giving one offspring. In the case of the matings to daughters then values should be used as proportions and multiplied. Therefore a sire with five litters to his own daughters which produced all normals in litters of 3, 4, 4, 5 and 6 would have a chance of error of:

$$0.71 \times 0.66 \times 0.66 \times 0.62 \times 0.59 \times 100 = 11.3\%$$

Thus after quite extensive matings the chance of being wrong is still quite high at 11.3%.

RANDOM MATING IN POPULATION

One of the problems with test mating as outlined above is that either known affected or carrier individuals have to be maintained for testing purposes (which may be costly and often impractical) or relatively high inbreeding (to own daughters) has to be undertaken. There is also the problem of what to do with individuals from such test matings which may be Nn in genotype and therefore of little use as breeding stock. Moreover testing has to be done for each defect separately.

There is great merit in having some recording scheme to evaluate matings in the population at large. If the incidence of a defect is known for the population then the number of progeny needed to 'exonerate' a sire mated at random can be assessed. At the same time as testing for a specific defect, other defects might also be located if the recording scheme is supported.

The probability of detection can be shown to be equal to:

$$1 - (1 - 0.5q)^n$$

which means that the chance of detection depends upon the frequency of the defect in the female population and the number of progeny. If the defect in question is lethal or if no females with it are in the population at large then the probability of detection becomes:

$$1 - [(1 + 0.5q)/(1 + q)]^n$$

Table 4.4 **Numbers of progeny to detect carrier status with random mating in the population (95% probability)**

Frequency of a allele in population	aa *females in population*	aa *females not in population*
0.50	11	17
0.30	18	24
0.10	58	64
0.01	598	604

For different genotype frequencies in the female population and for different degrees of detection the numbers of progeny needed in each case are given in Table 4.4.

For relatively rare alleles a considerable number of progeny will be required although these numbers would not be difficult to achieve with AI bulls which can produce many thousands of progeny. However, sires used in natural service cannot easily be assessed for very rare defects because such sires simply do not have enough progeny. Rare defects will be detected only fortuitously. It can, of course, be argued that if a defect is very rare then it presents no real problem to the population/breed under study.

Although it is desirable to seek to minimise deleterious recessives and other such defects, means to combat them must be kept in perspective. Breeding is about trying to produce desirable animals (whether this be efficient producers or those of desirable type) and these objectives should never be put in jeopardy by plans aimed at reducing rare defects, the effect of which on the population at large is, by definition, minimal.

5 Selection for Polygenic Traits

Introduction

So far emphasis has been mainly on simple traits controlled by one gene locus. Most economically important traits of farm livestock are not only controlled by many genes (polygenic) but they are influenced to a greater or lesser degree by environmental factors. In seeking to improve a character such as weaning weight of beef cattle there is (as yet) no way of knowing how many genes are involved, still less the genotype of individual animals. In order to bring about genetic progress in such a character there is a need to understand certain basic parameters of the population under study.

The assumption is usually made that you are working with a normal population which allows the qualities of normal curves to be applied to the data. If the data under study do not fit normal patterns then transformation of data into logarithms or some other system is usually required.

The animal breeder is not seeking to alter the nature of the normal population but to alter the mean of the population in either an upward or downward direction. There is information on the variable under study in terms of a series of measurements on individuals from the population. The mean can be calculated as well as the variance and standard deviation as described previously. Given these data and given populations which have certain relationships such as half-siblings or parents/offspring, it is possible to calculate certain parameters necessary to predict progress in changing the mean.

Heritability

The heritability expresses that part of the superiority of the parents which, on average, is passed on to the offspring. Heritability can be expressed as a percentage or proportion but ranges from 0 to 100% (0.0 to 1.0). The term h^2 is used to signify the heritability and it is known (see Chapter 5) that particular kinds of trait tend to be associated with certain heritability levels.

The phenotypic variation seen in a population is caused by a mixture of genetic and environmental variation. If σ^2 is used to signify variance with P, G or E indicating phenotypic, genetic and environmental variation then:

$$\sigma_P^2 = \sigma_G^2 + \sigma_E^2$$

Genetic variation can be broken down into three components: additive (A), dominant (D) and epistatic (I) while the environmental variation can be divided into general or permanent effects (E_g) and specific or common effects, i.e. common to members of a particular litter or to individuals from the same dam (E_S). This means that the above equation becomes:

$$\sigma_P^2 = \sigma_A^2 + \sigma_D^2 + \sigma_I^2 + \sigma_{E_g}^2 + \sigma_{E_s}^2$$

The additive part is of greatest interest since the heritability is defined as:

$$h^2 = \frac{\sigma_A^2}{\sigma_P^2}$$

Heritability is a ratio rather than an absolute figure, such that decreases in any of the other sources of variance will increase the heritability. Heritabilities are thus specific to particular populations although there is a tendency to find similar figures for the same character in different populations. In broad terms heritabilities were given in Chapter 1 (Tables 1.1 to 1.4) and in Table 5.1 some more specific figures are given, albeit not necessarily applicable to all populations. In most cases a range is given to encompass the figures normally seen in published literature.

Calculating heritabilities

Various methods exist to assess heritabilities using either regression or correlation techniques. If data for particular traits (e.g. growth rates) exist for both parents and offspring then regressions of

Table 5.1 **Examples of heritabilities in livestock (%)**

Cattle		Sheep		Pigs	
Calving interval	0–15	Lambs born	0–15	Litter size	0–10
Calves born	0–15	Lambs weaned	0–10	Pigs weaned	0– 7
Milk yield	30–40	Weaning weight	10–40	Weaning weight	0– 8
Fat yield	25–45	Fleece weight	30–40	Daily gain	21–40
Fat %	32–87	Staple length	30–60	Feed efficiency	20–48
SNF %	53–83	Fibre diameter	40–70	Killing out	26–40
Protein %	48–88	Crimps/cm	35–50	Fat depth C	62–65
Lactose %	28–62	Medullation	34–80	Fat depth K	42–73
Feed efficiency	40	Birth coat	59–80	Carcass length	40–87
Birth weight	38	Face cover	36–56	EM area	35–49
Weaning weight	20–50	Wrinkle	20–50	Leg length	46–50
Eye muscle area	40–70			Fillet weight	31–54
Daily gain	40				
Carcass lean	39				

offspring on either parent or offspring on mid-parent (the average value for sire and dam) can be carried out.

The regression of offspring on sire or offspring on dam gives a regression coefficient, the doubling of which gives the heritability. With offspring/mid-parent regressions the coefficient is equal to the heritability. Although these techniques are useful they can only be used if the character is measurable in both sexes. Thus egg/milk traits do not lend themselves to offspring/sire regressions. Moreover, dams are frequently selected animals which may bias matters as will maternal effects which bring about greater relationships between offspring of the same dam and thus inflate heritabilities.

Maternal effects are environmental effects, largely nutritional, which progeny have in common with their litter mates as a consequence of being carried in the same womb and then being suckled by the same dam. This relationship means that siblings not only show genetic similarities with each other and with their parents because they had their parents in common but environmental similarities due to the same or similar pre- and post-natal environments.

Of greater value are half-sib correlations where the study involves progeny from different sires (or dams) and in which measurements are made only on progeny and are not needed on the parent. These techniques are preferable to parent/offspring techniques because half siblings will usually have similar environments to one another and be measured at the same point in time. In contrast parents may be measured at a quite different point in time from their offspring and may also have been assessed under somewhat different environmental conditions to those in which progeny were measured. On the other hand, the correlation between half-sibs is quadrupled to give the heritability so that any errors in estimation are also quadrupled. Paternal half-sib correlations are preferable to those for maternal half-sibs largely because it is easier to get more of the former through the greater use of males and also because maternal effects are eliminated.

To be meaningful, heritability studies require considerable numbers of parents and offspring. Better estimates of paternal half-sibs are likely if there are large numbers of sires each with large numbers of progeny. A minimum of ten sires with at least five progeny each would be required, and ideally many more. It is better to have more sires than to have more progeny per sire once minimum numbers of progeny have been achieved.

Full-sibs are rarely used for heritability studies because maternal effects confound the data and the same is true of identical twin data. Identical twins are not easy to obtain but if used the differences between them reflect environmental effects. Heritabilities based on twins are usually much higher than those based on non-twin data and as a result twin and full-sib data are not of much practical value.

Methods of calculating heritabilities using examples are excellently described by Becker (1984).

A difficulty exists with threshold traits (see Chapter 3). In a single threshold situation animals either have or have not got the trait in question but the underlying genotype is unclear.

Calculation of heritabilities for threshold traits requires a population in which the incidence of the trait is measured and at the same time the incidence in a related group is needed. For example if cryptorchidism in a male pig population was at 5.0% and it was known that full sibs of affected pigs had an incidence of 15.0% the heritability of the trait could be assessed. The formula required is:

$$h^2 = t/r$$

where r is the degree of relationship and t is the correlation of liability to the trait in the two groups concerned. This t can be calculated from the formula:

$$t = \frac{X_P - X_R}{i}$$

where X_P and X_R are the distances in standard deviations of the threshold point from the mean for the population (P) and relatives (R) respectively, and i represents the distance in standard deviations of the mean of the affected population from the mean of the population from which it came. Values of i are read from Table 5.6 while values of X can be read from Table 5.2.

Returning to the cryptorchid pig example given above, an incidence of 5% gives an X_P value of 1.645 while that of 15% gives a value of X_R of 1.036 (both taken from Table 5.2). The value for i at 5% (read from Table 5.6) is 2.063.

$$\text{Thus } t = \frac{1.645 - 1.036}{2.063} = 0.295$$

Because full sibs have a relationship of 0.5 the heritability is given as:

$$h^2 = 0.295/0.5 = 0.59$$

The equation used here is a rough guide only and makes the assumption that variance is the same in affected and related

Table 5.2 **Deviation of the threshold point from the mean in a large population (value in standard deviations)***

Population affected (%)	X	Population affected (%)	X
0.50	2.576	8.0	1.405
0.75	2.432	10.0	1.282
1.0	2.236	15.0	1.036
2.0	2.054	20.0	0.842
3.0	1.881	30.0	0.524
4.0	1.751	40.0	0.253
5.0	1.645	50.0	0.000
6.0	1.555		

* A more extensive table is given by Falconer (1989).

individuals, which may not always be accurate. The result may underestimate the true figure to a slight degree.

Selection differential

In seeking to make progress it is not only important to know how inherited a character is, but also to assess what selection can be undertaken. Breeders make a conscious decision as to which animals will be used for breeding, depending upon the criteria being considered. In theory the breeder is seeking to select the 'best' or most appropriate animals as parents of the next generation and obviously differences of opinion may exist over such criteria. Even if there is unanimity about which is 'best' the selected animals may not survive to reproduce or may be less fertile than one hoped.

The selection differential is defined as the superiority of the animals selected to be parents over the mean of the population from which they came. The greater the selection differential then the greater the potential progress, other things being equal. Selection differential is largely determined by the breeder when selection is made but there are certain factors which control the limits to which a breeder can go.

Usually selection is more effective in males than females. This arises because fewer males are required than females and as a result greater selection intensity can be practised in males, especially if artificial insemination is widely used. An AI bull could, for example, mate 20 000 cows in a year whereas in natural service a bull might mate 40 or 50 cows. As a result many fewer males are needed in AI than is the case with natural service so that selection can be for the more outstanding animals.

There are, of course, some physical limitations. If a beef cow has six calves in her lifetime, of which half are daughters, then eventually one of those daughters will be needed to replace her. It may be that a particular cow does not have good enough daughters but for every such cow more daughters are needed from other cows if the herd size is to be maintained. A herd of 100 cows would give a final total of 300 daughters and therefore one third of these are needed as replacements. The breeder can do no better than retain the best 33% of the daughters. In contrast a beef bull in natural service might produce, say, 50 sons so that selection can be made at the level of the best one in 50. With AI as was seen above prospects would be even better.

Breeders decide on (select) the sires and dams of the next generation. Since fewer males are required greater selectivity is possible in choice of their fathers and mothers. Daily farmers, for example, can be much more selective about which cows are good enough to breed them a bull and much less about which cows will be used to produce daughters. They cannot guarantee which cows will give them sons but they can determine from which cows sons would be acceptable. There are thus four routes to progress in selectional terms:

> Males   to breed   Males
> Females to breed   Males
> Males   to breed   Females
> Females to breed   Females

In dairy cattle some 75% of genetic progress results from the choice of the parents of bulls and only around 5% from the selection of mothers of future females. New multiple ovulation/embryo transfer techniques (MOET) may increase progress by the female route (see Chapter 9).

The more animals measured and the fewer that are needed as parents, the greater the intensity of selection that is feasible. In a sheep flock a high reproductive rate coupled with good survival rates means that more sheep are available for selection. Similarly, the greater the variability in the population the easier it may be to identify and select animals that are at the extreme ends of the performance scale.

Imagine a beef herd in which post-weaning daily gain is being selected. Assume average daily gain for bulls is 1.5 kg and that for heifers is 1.3 kg. If only bulls averaging 2 kg are used for breeding while heifers averaging 1.6 kg are used for 'bull mothers and those averaging 1.5 kg are used to produce females the selection differentials would be as shown in Table 5.3.

If no selection were undertaken in females but the same intensity practised in males then the total selection differential for the four routes would be 0.50 + 0 + 0.50 + 0 = 1.00 and the mean selection differential would be + 0.25 kg. Failure to undertake any selection on the female side would reduce the potential progress.

These are the predicted selection differentials rather than actuals. It is rare for selected males to produce equal numbers of offspring

Table 5.3   **Selection differentials**

Bulls to breed bulls (2.00–1.50)	= +0.50 kg
Heifers to breed bulls (1.60–1.30)	= +0.30 kg
Bulls to breed heifers (2.00–1.50)	= +0.50 kg
Heifers to breed heifers (1.50–1.30)	= +0.20 kg
Total	= +1.50 kg
Mean selection differential (1.5/4)	= +0.375 kg

Table 5.4   **Calculation of a Selection differential**

Bull	Gain (kg)	Progeny measured	Deviation of bull	Weighted deviation
1	2.30	15	+0.80	+12.00
2	2.20	19	+0.70	+13.30
3	2.00	25	+0.50	+12.50
4	1.80	10	+0.30	+ 3.00
5	1.70	11	+0.20	+ 2.20
Totals		80	+2.50	+43.00
Average			+0.50	+ 0.538

even if mated to the same numbers of females and this is certainly true in litter producers where litters vary in size. The actual selection differential is that obtained after weighting each parent's superiority by the numbers of progeny born and measured.

Suppose five bulls were selected for breeding from a population in which post-weaning gain averaged 1.5 kg. Further suppose that these five bulls had gains of 2.3, 2.2, 2, 1.8 and 1.7 kg respectively. Their average gain is 2 kg which is a mean of +0.5 kg expressed as a deviation. When numbers of progeny are considered the situation may be as shown in Table 5.4.

In this instance the actual selection differential achieved is slightly higher than that anticipated because the better bulls tended to have slightly more progeny than the poorer ones.

The phenotypic standard deviation of a character is the way in which variation is usually described and if the standard deviation is expressed as a percentage of the mean it becomes the coefficient of variation (CV). Some characters like milk yield are quite variable (high coefficients of variation) while others like daily gain are less

Table 5.5 **Phenotypic standard deviations in farm livestock***

Species	Trait	σ	units
Dairy cattle	Milk yield	250	kg
	Milkfat yield	40	kg
	Milkfat percentage	0.49	%
	Lactation length	30	days
	ICC milk**	125	kg
	ICC fat yield**	4.7	kg
	ICC protein yield**	3.8	kg
	CGI**	96	units
Beef cattle	Birth weight	6	kg
	200-day weight	25	kg
	400-day weight	30	kg
	pre-weaning gain	0.13	kg/day
	post-weaning gain		
	feedlot	0.10	kg/day
	pasture	0.08	kg/day
Sheep	Lambs reared	0.6	lambs
	Lamb weaning weight	3.6	kg
	Body weight (hogget)	4.5	kg
	Ewe fleece weight	0.5	kg
Pig	Daily gain	0.06	kg/day
	Feed conversion	0.20	feed/gain
	Backfat (C)	2.5	mm
	Backfat (K)	2.7	mm
	Backfat (S)	4.0	mm
	Carcass weight	1.3	kg
	Dressing percentage	1.60	%
	Litter size	2.8	piglets

* Figures are guides only. Specific populations may vary markedly from these.
** Holstein/Friesians only.

Table 5.6 **Selection intensity (infinite population size)**

Percentage selected	Mean superiority (in standard deviations)
1	2.665
2	2.421
3	2.268
4	2.154
5	2.063
10	1.755
15	1.554
20	1.400
30	1.159
40	0.966
50	0.798
60	0.644
70	0.497
80	0.350
90	0.195

so. Table 5.5 shows some examples of phenotypic standard deviations for certain traits of farm livestock.

Selection intensity can be defined in terms of standard deviations thus:

$$\text{Intensity (i)} = \frac{\text{Selection differential}}{\text{Phenotypic standard deviation}}$$

In animal breeding you are working with normal populations with established properties, so it is possible to estimate the superiority in standard deviations of any given proportion of the population. Such tables are quite extensive and are given by Pearson (1931) and Becker (1984). A very brief extract is given in Table 5.6.

With small population sizes the values would differ from those in Table 5.4 but those given will be valid enough for populations of 300 to 400 individuals or more. Given tables such as these, predictions of progress can be made for specific selectional standards. Values appropriate for small populations appear in Becker (1984) in considerable detail.

Table 5.7 **Female selection intensity in sheep**

Lamb weaned/100 ewes mated	80	100	120	140	160	180	200
Females available	40	50	60	70	80	90	100
Replacements needed*	18	18	18	18	18	18	18
Per cent retained†	45	36	30	26	23	20	18
Selection intensity (i) ‡	0.88	1.04	1.16	1.25	1.32	1.40	1.46
Relative progress (column 1 = 100)	100	118	132	142	150	159	166

* Includes small addition to allow for deaths.
† Per cent retained of total available.
‡ Figures taken from larger version of Table 5.6.

The importance of fertility in aiding selection is shown in Table 5.7 using sheep with different lambing percentages but assuming equal sex distribution and a six-year life in the flock.

As lambing percentage increases more sheep are available for selection and since a six crop life is assumed constant, selection intensity increases with fertility. Relatively speaking the 200 lamb crop flock (feasible in lowland areas in Britain) is 66% more effective in selection terms than would be the case in flock with an 80% crop, other things being equal. If the actual length of life were less than six crops then the advantage of superior lambing percentage becomes even more apparent.

Generation interval

Different species reproduce at different rates. They reach breeding age at different points and they have differing gestation lengths. The decision 'when to breed' is partly determined by physiological considerations but also by man. He can decide to calve down a heifer at two years of age or at three depending upon economic considerations. This type of decision also has genetic implications as the earlier animals are bred from the quicker one arrives at the next generation which, hopefully, will be superior in genetic terms to the present one.

Generation interval is defined as the average age of parents when their progeny are born. In effect it defines the length of time between generations. This will vary with particular populations and between species. Puberty will set a lower limit for generation interval but the average value will usually be higher. In man, although birth is feasible at 16 years or even less, the mean generation interval is closer to 25 years. Values for farm livestock species are shown below (in years):

Horse	9–13	Dog	4–5
Dairy cattle	5–7	Sheep	3–4
Beef cattle	4–5	Pig	1–2

Generally, a breeder is interested in progress per unit of time (usually per year) and generation interval allows such a calculation. The quicker the generation turnover the more rapid the progress, provided that a quick turnover is consistent with still having data on the parents being selected. In traits measured on the live animal (e.g. daily gain) generation interval can be kept relatively low but in dairy cattle where bulls have to be assessed on progeny the generation interval is increased by waiting for progeny records.

Genetic gain

The three components of progress – heritability, selection differential and generation interval – have now been defined. They are combined to measure progress thus:

$$\text{Genetic gain/generation} = \text{heritability} \times \text{selection differential}$$

$$\text{Genetic gain/year} = \frac{\text{heritability} \times \text{selection differential}}{\text{generation interval}}$$

These equations are at the core of all improvement. Maximum gain will result when the heritability and SD are high and the GI low. If the heritability is low then a high SD and low GI are needed. Anything that increases the heritability and/or the selection differential aids progress, as does anything which reduces the generation interval, provided these actions do not affect other items among the three.

Measurement of genetic gain

Long-term experiments with farm livestock are not common, especially in terms of generations. In species like mice and *drosophila* trials have lasted 30 generations (about 180 years in cattle terms!) and the divergence between upward and downward selection has been as much as 20 standard deviations of difference. Similar progress might be feasible in farm livestock given long enough selection.

It is necessary to measure progress since a simple elevation of the mean may indicate environmental change rather than genetic advance. There are several ways in which genetic progress might be assessed. One way is to set up control herds/flocks/lines in which selection is purely random and selected lines can then be compared with this control population. If maintained in the same environmental conditions the difference between control and selection lines will measure genetic progress.

Control populations need to be large enough to allow them to be closed without inbreeding levels becoming too high. This means that larger numbers of males than would be needed in a commercial unit are used and family sizes are best kept constant. Such a control unit was set up in pigs at Cockle Park (Newcastle University) comprising 16 boars and 32 sows. Each sow was replaced by a daughter and each boar by a son selected a random. Eventually such control units become too far behind selection lines to allow meaningful comparisons and to avoid random drift they are usually reconstituted. The Cockle Park line ran for about 14 years before being disbanded.

Another type of control would be to store semem from sires of proven merit and to use this at intervals in the population at large. Every five to ten years might be feasible. Such a procedure is easier in cattle than pigs or sheep because of the greater ease of freezing semen. Frozen embryos might also be used to produce similar results, but would be more expensive.

In any selection work limits might be arrived at when the selected criterion reach a plateau having first starting to slow down. Usually this means that additive variation has been used up, but if new material is introduced at this point it may be feasible to advance again. New material could be from a different population or breed, or by introducing mutations, or by genetic engineering. The theoretical consequences are shown in Fig. 5.1.

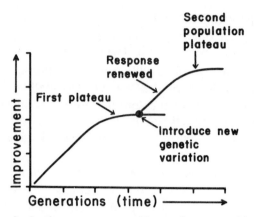

Fig. 5.1   **Renewed selection response reaching a plateau at a higher level**

Agreement between estimated and realised parameters

In this chapter genetic parameters such as heritability, standard deviation and generation interval have been discussed. In Chapter 7 genetic and phenotypic correlations between traits are discussed. It has been stressed that these parameters relate only to the population from which they were derived and great care has to be taken not to overemphasise the importance of these parameters. In deriving parameters various assumptions are made, such as the population being unselected, normal, large and randomly mated. It is usually further assumed that environmental effects are randomly distributed and that there are no correlations between genotype and environment. These assumptions are quite sweeping and may not always be valid.

In a comprehensive review of animal breeding literature Sheridan (1988) concluded that in laboratory species the realised heritabilities were as frequently under- as overestimated but that in commercially important livestock realised heritabilities were frequently overestimated. This could result in breeding schemes being adopted that were not fully justified. This emphasises the desirability of estimating parameters from selected lines in preference to base populations and in not accepting genetic parameters as being valid for too long a period.

Whatever the care taken in assessing genetic parameters and in predicting selection response it must not be forgotten that we *are* dealing with estimates and predictions which by their very definition cannot be exact and precise. Most genetic progress will be below that which was anticipated. This will stem from overestimation of genetic parameters but will also result from the fact that selected animals will not always perform in practice as predicted in theory and that selection aims may not turn out to be quite what was envisaged.

6 Aids to Selection

Pedigree information

A pedigree is simply a record of ancestry and is not synonymous with purebred since a crossbred animal might just as easily have a record of its ancestry. Often a pedigree, even an official one issued by a breed organisation, is little more than a list of names or numbers and thus has only limited value in terms of predictive worth. Some purebred pedigree breeders argue that they memorise the qualities of the animals in the pedigree but such reliance on memory is fallible. Pedigrees which contain information on performance data of the ancestors are of greater value than those with mere names but even then pedigree data must not be overemphasised because, as will be seen later, the predictive value of a pedigree is limited.

The basis of a pedigree is the relationships it carries. A four or five generation pedigree is normally as much as one expects to see and these will contain 30 and 62 ancestors respectively. The breeder's problem is to determine what consideration or 'weighting' to give to the various ancestors.

Figure 6.1 shows the pedigree of a Shorthorn bull Roan Gauntlet from the early development days of the breed. It actually shows some inbreeding but its purpose here is to illustrate pedigree features.

The pedigree in Fig. 6.1 is a four generation one, the subject's generation not being counted. It is normal policy to list sires above and dams below in each generation pairing. What is called the tail female line is that going from dam to dam, i.e. Princess Royal—Carmine—Cressida—Clipper, while the tail male line would be Royal Duke of Gloucester—Grand Duke of Gloucester—Champion of England—Lancaster Comet.

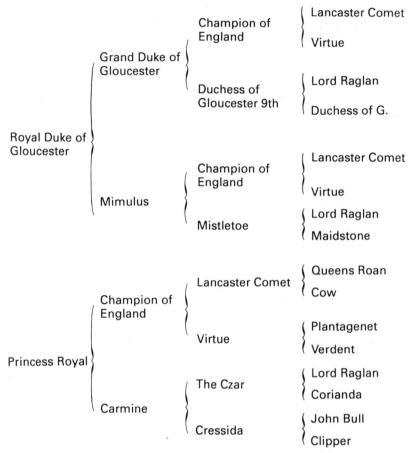

Fig. 6.1 **Pedigree of Roan Gauntlet (Shorthorn)**

Any animal in each generation derives half its genetic make-up from each parent and that is valid in every case. In theory each animal thus derives a quarter of its genetic make-up from each grandparent but this is not true in each instance. Roan Gauntlet would get half his genes from Royal Duke but it is not certain that half of those came from Grand Duke. The further back you go the less influence ancestors will have. In generation six there are 64 ancestors, but as a bull has only 60 chromosomes some of these ancestors either appear in name only or a mere fraction of a chromosome exists in the subject from some of the ancestors.

There is no evidence that particular features stem from sires or from dams though clearly an individual may obtain a 'better' sample of genes from one parent than the other. Ideas such as the sire giving size and the dam colour are without foundation although, in any particular instance, an individual may show the size of its sire or the colour of its dam.

The nomenclature system in many species, especially cattle, can be misleading. If a particular 'female' family started with a cow called Pearl, then all female descendants in tail female line might be called Pearl with a number. Pearl 144th would be the 144th Pearl born into the family and trace back in tail female line to the original Pearl. Too much meaning should not be read into this system. Pearl 144th may be a paternal half sib to Buttercup 106th and have more relationship to that cow from a different 'family' than to any member of the Pearl 'family'. Families of this type are not very meaningful in genetic terms, and family selection in the genetic sense has no relationship to these so-called female families. Similarly, the fact that all animals from a particular herd/flock/stud carry the same affix/prefix does not necessarily make them a 'line' or 'strain' or even indicate any relationship.

The value of pedigrees depends upon the closeness of the relationship to the animal under study and the heritability of the trait. The higher the heritability the greater the value of pedigree data, but also the more important information on the subject's own performance becomes. Table 6.1 shows the relative efficiency of pedigree data at different heritability levels. Efficiency relates to the correlation between breeding values or genetic worth and the phenotypic measure being studied. In all cases single records are assumed. By single record it means one lactation yield per animal or one weaning weight as opposed to multiple records.

In any breeding programme a potential pedigree can be constructed before the mating is even made and predictions are possible. Official breed society pedigrees exist as a record of ancestry but evidence suggests that between 5% and 10% of pedigrees may be inaccurate through mistakes being made, however innocently, about parentage. The more extensive the system of management the greater the risk of such errors. More errors are thus likely in an extensive beef ranch using natural service bulls than in a small dairy herd using AI.

Table 6.1 **Relative efficiency of pedigree records (%)**

Heritability (%)	Pedigree records on the			Subject's own performance
	Parents only	Parents plus Grandparents	Complete pedigree	
10	22	27	29	32
20	32	37	39	45
30	39	43	45	55
40	45	49	50	63
50	50	53	54	71
60	55	57	57	77
70	59	61	61	84
80	63	64	64	89

Performance testing

Performance testing involves recording the performance of the animal under study and making decisions in the light of those data. A geneticist would term this individual or mass selection but breeders would understand performance testing better.

The usefulness of performance testing depends upon the degree to which superiority of an individual compared with its contemporaries acts as a guide to the future performance of the progeny of that individual. The performance test seeks to be a guide to the breeding value of the individual and this can only be assessed if individuals are compared with others treated in the same way at the same time.

To be effective, performance tests need to be strictly controlled to ensure that the animals being tested *are* being treated in the same way. If, for example, beef bulls are to have their post-weaning growth measured from, say, 200- through to 400-days of age, then the bulls need to be located in a particular unit and fed the same diets post-weaning. The degree to which there is equal treatment will determine the validity of the testing. However, pre-test environment could still influence the post-weaning tests. If the bulls came from mothers of different ages/parities then this might influence post-weaning performance. Similarly bulls treated exceptionally well prior to the test may not perform as well on the test as those given much poorer pre-test treatment but which are then 'compensating' on test.

Performance tests are really only valid for individuals tested at the same time and place on the same regime though some latitude may exist as to the meaning of 'same time'. Ideally animals should start at the same time and be kept together but contemporaneity might extend to animals starting the test over a particular range in time. Testing at special centres is easier than on-farm testing, but more costly. The difficulty with on-farm testing might be obtaining enough animals to test at the same time. Moreover tests on one farm might not be comparable to those on another just as tests on a farm at one time may not be comparable to tests on the same farm at a different time.

Breeding value could be defined as the heritability of the character × the deviation of the individual's performance from the mean of its contemporaries' performance. The higher the heritability the greater the potential for performance testing (see Table 6.1). Reliability of testing is best expressed by looking at subsequent progeny from performance tested animals. If the ranking order of bulls on a performance test was more or less identical to their ranking order on the basis of their progeny then the test will be effective.

In general terms, performance testing is only suitable for traits of moderate to high heritability and is of no use if the characteristic cannot be assessed in the live animal. Dairy bulls are not performance tested other than perhaps for growth, but beef bulls, sheep and pigs are frequently tested on their own performance. In pigs, testing can be effectively carried out for some carcass traits without having to slaughter the animal; increasingly this is becoming true of cattle and sheep. Measuring the ability of an animal to do something, e.g. racing performance of a horse or herding ability of a sheepdog would also be examples of performance testing. It has to be realised that measurement itself can add variability to the character under study. The more objective and uniform the way measurement is made, the more valuable will be that measurement. Subjective measurements can, nevertheless, be valuable in making performance tests.

Multiple records

Performance testing usually relates to one record as a bull can only be assessed once on its growth from 200 to 400 days. Some animals can, however, be assessed more than once. Fleece weight can be assessed each year, as can lactation yield or litter size. Good animals ought to be able to perform well each time and the extent to which a record can be repeated is termed the repeatability (R). This is defined as:

$$R = \frac{\text{Genetic variance} + \text{General environmental variance}}{\text{Total phenotypic variance}}$$

or

$$R = \frac{\sigma_A^2 + \sigma_D^2 + \sigma_I^2 + \sigma_{E_g}^2}{\sigma_A^2 + \sigma_D^2 + \sigma_I^2 + \sigma_{E_g}^2 + \sigma_{E_s}^2}$$

The numerator and denominator in this equation are identical except for the specific environmental effect. Characters which have a high 'specific' effect will have a low repeatability and those with a low 'specific' effect will be highly repeatable. Repeatability indicates the correlation between records and it also sets an upper limit on the heritability (see Chapter 5) since the heritability is lower than or, very rarely, equal to, but is never higher than the repeatability.

By taking repeat measurements the breeder is reducing the importance of the specific environmental effect (i.e. those peculiar to a particular record) and this will increase the heritability of the trait. The heritability is increased by multiplying by the following formula:

$$\frac{n}{1 + (n - 1)\,R}$$

where n is the number of records and R is the repeatability. Table 6.2 shows this formula worked out for different numbers of records and repeatabilities.

If, for example, you are looking at a character with a heritability of 0.30 and a repeatability of 0.5 then by having two records available the heritability increases by 1.33 times and becomes 0.40; with three records it increases to 0.45 and with six records it reaches 0.51.

Table 6.2 illustrates the advantages of waiting for more records and generally speaking the value of waiting decreases as the repeatability increases. A high repeatability means that one record is

Table 6.2 **Confidence factor of heritabilities using several records**

Number of records	Repeatability								
	0.1	0.2	0.3	0.4	0.5	0.6	0.7	0.8	0.9
1	1.00	1.00	1.00	1.00	1.00	1.00	1.00	1.00	1.00
2	1.82	1.67	1.54	1.43	1.33	1.25	1.18	1.11	1.05
3	2.50	2.14	1.88	1.67	1.50	1.36	1.25	1.15	1.07
4	3.08	2.50	2.11	1.82	1.60	1.43	1.29	1.18	1.08
5	3.57	2.78	2.27	1.92	1.67	1.47	1.32	1.19	1.09
6	4.00	3.00	2.40	2.00	1.71	1.50	1.33	1.20	1.09

a good guide to future ones and hence waiting is hardly needed. In contrast low repeatability characters benefit from additional records.

Lifetime records are useful in identifying animals that have proved their ability to survive and perform. However, the need to wait for more data not only increases the heritability but also the generation interval and overall this may produce a slower rate of genetic progress.

Having access to multiple records allows an individual's breeding

Table 6.3 **Breeding value assessments for weaning weight of beef cattle (hypothetical example)**

Calving	Calf weight as deviation from herd mean (kg)			
	Cow A	Cow B	Cow C	Cow D
1	−20	+ 50	−10	+ 35
2		+ 30	+15	+ 30
3		+ 40	+25	+ 35
4			+20	+ 40
Total deviation	−20	+120	+50	+140
Mean deviation	−20	+ 40	+12.5	+ 35
Reliability	0.30	0.45	0.48	0.48
Breeding Value	− 6	+ 18	+ 6	+ 16.8

value to be assessed. The figures in Table 6.2 are multiplied by the heritability and then by the deviation of the animal's performance from contemporaries. In each case deviations are assessed for each record and summed. This is shown in Table 6.3 for weaning weight in beef cattle and assumes a heritability of 0.3 and a repeatability of 0.5.

The order of merit would be cows B, D, C and A, and certainly B and D look to be above average whereas C is not particularly

Table 6.4 **Repeatability of farm livestock traits (as %)**

Species	Trait	Repeatability
Dairy cattle	Services per conception	12
	Annual non-return rate	6
	Bull ejaculate volume	70–80
	Milk yield	40–60
	Milkfat percentage	40–70
	Milking rate	80
Beef cattle	Birth weight	20–30
	Weaning weight	30–55
	Yearling weight	25
	Daily gain to weaning	7–10
	Body measurements	70–90
Sheep	Ovulation rate	60–80
	Lambs born/ewe mated	15
	Lambs born/ewe lambing	30–40
	Lambs weaned/ewe mated	18–20
	Birth weight	30–40
	Lamb gain	38–50
	Fleece weight	30–40
	Various wool traits	50–80
Pigs	Litter size/birth	10–20
	Litter size/weaning	10
	Litter weight/birth	25–40
	Litter weight/8 weeks	5–15
	Birth weight/pig	20–40
	Weaning weight/pig	10–15
	Adult live weight	35
	Eye muscle area	95

outstanding. Cow A may have had some problem in the case of her only calf and may or may not repeat this sort of below average performance. Breeding values are only predictions not exact features but the more information that they are based upon the more accurate are they likely to become.

Some measures of repeatability are shown in Table 6.4.

Progeny testing

Decisions can be based upon the performance of progeny rather than the performance of the subject. Usually this is applied to males because it is easier to obtain more progeny from males than females.

Progeny represent a sample of the genes of their father (and mother) and, if enough samples are drawn, these will give a reasonable idea of the superiority (or otherwise) of that sire. This is, of course, only possible if the progeny of any particular sire are compared with the progeny of other sires in the same environmental conditions. The performance of a bull tells one what he *might* do if used, whereas a progeny test tells one what he *is* doing.

Progeny testing is time consuming and increases the generation interval. For example, a dairy bull will be about six years of age before he has sufficient progeny lactating to allow any estimate of his breeding worth for milking traits. For this reason performance testing is generally preferable but it obviously cannot be undertaken for milk yield in bulls.

Usually progeny testing is undertaken for traits that have a low heritability and for which performance results are thus a poor guide. It is also used for traits expressed only in one sex (e.g. milk production) and those expressed only after death (e.g. carcass composition).

To be effective progeny tests require:

(1) As many sires on test as possible (5–10 minimum).
(2) The random mating of dams to ensure that certain sires are not mated simply to very good or very poor dams and that sires get similar numbers of young/old mates.
(3) As many progeny per sire are produced as possible (ten minimum for some traits but several hundred for others).
(4) Progeny must be unselected during the test.

Table 6.5 **Relative efficiency of progeny records (%)**

Heritability (%)	Number of progeny tested							
	5	10	20	40	60	80	100	120
10	34	45	58	71	78	82	85	87
20	46	59	72	82	87	90	92	93
30	56	70	79	87	91	93	94	95
40	60	73	83	90	93	95	96	96
50	65	77	86	92	95	96	97	97
60	68	80	88	94	96	97	98	98
70	72	82	90	95	96	97	98	98
80	75	84	91	95	97	98	98	98

(5) Progeny must be treated in the same way or comparions made within herd/flock and year/season groupings.

Given these standards progeny testing can be very effective. The reliability of testing increases with the heritability and the number of progeny in the test. This is shown in Table 6.5 which can be compared with Table 6.1 to indicate the advantages of progeny testing over performance and pedigree information.

In general, progeny testing gives more information than any other source of data with a rapid increase in relative accuracy as progeny numbers increase, but little additional accuracy once about 40 progeny are obtained. Progeny give more information than the corresponding numbers of full or half-siblings (compare Tables 6.5 and 6.6) and considerably more than pedigree data. At low heritabilities (10% or 20%) five progeny are as useful as performance data and at low heritabilities only progeny testing on sufficient numbers of progeny offers any reasonable accuracy.

Information from other relatives

Sometimes it is possible to obtain data from relatives such as full-siblings (brothers and sisters) or from half-siblings (usually by the same sire). The relative accuracy of these records is shown in Table 6.6 which can be compared with Tables 6.1 and 6.5.

Table 6.6 **Relative efficiency of sibling information (%)**

Heritability (%)	Number of full sibs			Number of half sibs			
	2	4	8	5	10	20	40
10	22	29	38	17	23	29	36
20	30	39	48	23	29	36	41
30	36	45	54	27	33	39	44
40	41	50	58	30	36	41	45
50	45	53	60	32	38	43	46
60	48	56	62	34	40	44	47
70	51	58	63	36	41	45	48
80	53	60	65	37	42	46	48

Sibling selection tends to be most used in dairy cattle, pigs and poultry, and in general terms siblings very quickly become as, or more, useful than pedigree data. Half-siblings are less useful than full-siblings but at low heritabilities can be of value and, of course, half-sibling data may be more readily obtained than full-sibling information, e.g. in dairy cattle records.

Family selection is often confused because of the terminology used by cattle and pig breeders (see Chapter 6) but in genetic terms families usually relate to full or half-sib families. Sire families would be progeny of a given sire out of different dams while dam families would be progeny of a given dam by different sires. These would be half-sib families while full-sib families would have the same sire and dam.

Falconer (1989) has looked at various forms of selection which are illustrated in Fig. 6.2.

Individual selection involves taking the best individuals regardless of family, while family selection involves selection of the best families. Within-family selection involves selecting a proportion from each family and this highlights a major difference between family and within-family selection. Using family selection some families are completely discarded and inbreeding risks might increase as fewer families remain. Example D in Fig. 6.2 illustrates the case where within-family selection is most useful. Here there are very large differences between families but minimal variation within.

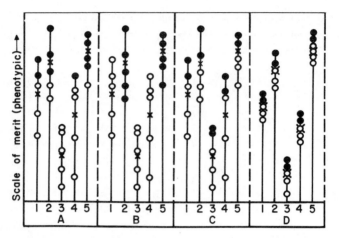

Fig. 6.2　**Different methods of family selection (redrawn from Falconer)**

Family selection is most useful when the genetic relationship between members is large and the phenotypic relationship is small. This is best seen in inbred lines.

Practical comparisons

If the facility to test 1000 pigs is available then you could either use this to performance test 1000 males, or alternatively progeny test differing numbers of sires with varying progeny numbers. Table 6.7 shows testing of 100 sires each with ten progeny, or 200 sires each with five progeny. A heritability of 30% is assumed. It is necessary to retain 50 males for breeding. The figures in the final column of Table 6.7 are the relative rates of progress. Progeny testing is more

Table 6.7　**Progeny v. performance testing**

Testing system	Number of boars	Proportion retained	Selection differential (S)	Relative accuracy (RA)	S × RA
Performance	1000	0.05	2.063	0.55	1.13
Progeny	100	0.50	0.798	0.70	0.56
Progeny	200	0.25	1.271	0.56	0.71

Table 6.8 **Number of bulls v. reliability of the test**

Bulls tested	16	20	25	40	80	100	166
Progeny/bull	125	100	80	50	25	20	12
Per cent kept	63	50	40	25	13	10	6
Superiority (S)	0.60	0.80	0.97	1.27	1.63	1.76	1.99
Reliability (R)	0.95	0.94	0.93	0.90	0.82	0.79	0.70
S × R	0.57	0.75	0.90	1.14	1.34	1.39	1.39

accurate (higher RA) but leads to lower selection differentials. Moreover, in performance testing you are dealing with the standard deviations for individuals whereas the standard deviations for groups of progeny will be reduced by $\sqrt{n}$ where n is the number in a group. Additionally there will be a greater generation interval attached to progeny testing. Performance testing would be preferable when dealing with a character as highly inherited as 30%.

Another way of using these figures would be in relation to progeny testing of dairy sires. Assume a heritability of 0.30 and a total of 2000 milk recorded cows available with ten progeny tested sires required each year. You could use different numbers of progeny per bull and thus test different numbers of sires. Some examples are shown in Table 6.8.

Although a test on a mere 20 progeny may not appear very useful, the greater selection pressure available by selecting from a larger sample of bulls outweighs this. In terms of effectiveness it is better, in this instance, to test around 100 bulls with 20 daughters each, than to test fewer bulls more efficiently, but to retain a higher percentage of those tested.

7 Selection for Multiple Objectives

Introduction

So far breeding and selection have been described in terms of single characters. In practice, few breeders select for only one thing. Livestock breeders usually have a series of objectives which will include specific production traits as well as, perhaps, some aspects of physical appearance. The more things you seek to obtain the harder it will be to achieve them all, and breeding is best directed towards what is feasible in any given situation. Selection for several objectives involves knowledge of certain genetic parameters if success is to be achieved.

Relationship between characters

Breeders are aware that certain characters seem to be related to others. Selection for increased wither height would be likely to lead to increased body weight. Selection for milk yield is likely to reduce the milkfat content of that yield. Breeding for increased feed efficiency in meat animals will probably be associated with leaner carcasses.

In some cases the beliefs about relationships between traits may be based upon ideas passed down from generation to generation and these are not always valid. Pronounced milk veins on an udder do not necessarily imply high milk yield any more than a thin skin denotes high milk potential. Some beliefs about relationships are without foundation but others have a basis in fact and a biological explanation.

Valid relationships stem from physiological ones, and it appears that the genes which affect a particular character may also affect other

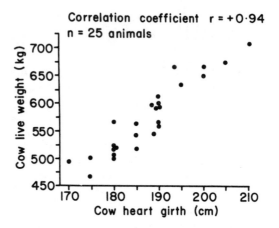

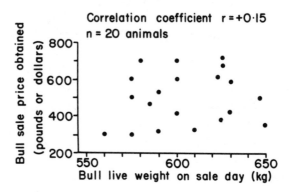

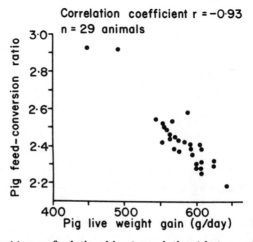

Fig. 7.1   Different types of relationships (correlations) between traits

characters, just as environments which affect one character in a specific way may affect others. Relationships between traits are called correlations and these can range from -1.0 through to $+1.0$ with the former indicating high negative relationship and the latter high positive relationship. Different kinds of relationships are shown in Fig. 7.1.

If two characters are measured in a series of individuals then the correlation between these characters is termed the *phenotypic correlation* (r_P). This phenotypic correlation can be divided into genetic and environmental parts, termed r_G and r_E respectively. Genetic correlations exist because the same genes affect both characters (a phenomenon termed *pleiotropy*) while environmental correlations exist because particular environmental conditions produce certain features in the animals, which result in a correlation between these features and the environmental conditions.

In mathematical terms these correlations can be described as:

$$r_P = r_G + r_E$$

Genetic correlations can be useful to a breeder. If two characters are related, the consequence of increasing one upon response in the other can be assessed. Selecting bulls for larger 400-day weights would be quite successful because the character is quite highly inherited. However, an increase in 400-day weight would involve some increase in weight at other ages because of positive genetic correlations. Birth weight might well increase and greater dystocia might result. Knowing this the breeder can select heavier 400-day weights but use bulls which are known to give fewer problems at

Table 7.1 **Correlations among traits in dairy cattle**

Trait	MY	BY	PY	FP	PP
Milk yield (MY)		0.82	0.87	−0.27	−0.18
Butterfat yield (BY)	0.88		0.86	0.26	−0.11
Protein yield (PY)	0.95	0.93		0.04	0.22
Fat % (FP)	−0.20	0.24	−0.01		0.55
Protein % (PP)	−0.19	−0.04	0.06	0.49	

Source: Gibson (1987) who drew from various sources.

calving. This latter aspect requires large scale progeny testing to evaluate.

Then again, knowing that characters are related, there may be an opportunity to select for one character and bring about a related or correlated response in a second character which might be much harder or more costly to measure. Some correlations between traits are given in Tables 7.1 to 7.4. In these tables genetic correlations appear above and phenotypic correlations below the diagonal. In looking at these tables it must be realised that these represent samples from the literature. Correlations, particularly phenotypic ones will not only depend upon which breed is being assessed in which location, but also upon the type of diets. For example pigs fed on *ad libitum* diets may show different relationships between traits than those fed under restricted conditions.

Table 7.2   **Correlations among traits in beef cattle**

Trait	BW	GR	FE	CF	KO	CLP
Birth weight (BW)		0.40	0.10	N/A	N/A	0.05
Growth rate (GR)	0.30		0.28	0.25	N/A	−0.22
Feed efficiency (FE)	0.10	0.39		−0.14	N/A	N/A
Carcass fatness (CF)	0.12	0.30	0.06		N/A	−0.60
Killing out % (KO)	0.04	0.45	0.20	0.39		N/A
Carcass lean prop (CLP)	0.08	−0.25	−0.06	−0.57	−0.56	

Source: Simm *et al.* (1986) and Preston & Willis (1976) who drew from various sources.
N/A = not available.

Table 7.3   **Correlations among traits in Merino sheep**

Trait	BW	GFW	FD	SL	C
Body weight (BW)		0.26	0.12	0.04	0.40
Greasy fleece wt (GFW)	0.36		0.19	0.70	−0.20
Fibre diameter (FD)	0.13	0.13		0.44	−0.10
Staple length (SL)	0.10	0.30	0.11		−0.34
Crimps/inch (C)	0.05	−0.21	−0.13	−0.22	

Source: highest values of Newton Turner & Young (1969) who drew from various sources.

Table 7.4 **Correlations among traits in pigs**

Trait	DG	FE	KO	CL	BF	EMA
Daily gain (DG)		−0.76	−0.19	0.14	−0.15	−0.11
Feed efficiency (FE)	−0.73		0.01	−0.08	0.21	−0.34
Killing out % (KO)	−0.17	−0.05		−0.40	0.28	0.36
Carcass length (CL)	0.07	−0.04	−0.19		−0.30	−0.08
Backfat thickness (BF)	−0.07	0.19	0.19	−0.22		−0.28
Eye muscle area (EMA)	−0.03	−0.16	0.15	−0.05	−0.13	

Source: Dalton (1985).

Although phenotypic correlations are readily estimated the calculation of genetic correlations is more difficult and usually requires not only related groups of animals (e.g. a series of paternal half-sib family groups) but large numbers. The fact that phenotypic correlations are positive and high gives no clue to underlying genetic correlations and selection is hampered if positive phenotypic correlations are masking negative genetic correlations.

It should be appreciated that the data given in Tables 7.1 to 7.4 are little more than illustrative guides and alternative results will ensue from different populations. Care must thus be taken in using the data presented or in reading too much into them.

Usually selection is most successful if made directly for the character in which you are interested. However there are occasions when selection for a correlated character can bring about greater progress in the trait concerned than direct selection. If you look at direct response to selection for trait A(R) and correlated response in A by selecting for trait B(CR) then the consequences will depend upon the respective heritabilities of traits A and B, the intensity of selection possible for A and B and the genetic correlation between them. This is expressed:

$$\frac{CR_A}{R_A} = r_G \frac{i_B}{i_A} \sqrt{\frac{h_B^2}{h_A^2}}$$

Suppose selection was for litter size (h^2 0.10) or for yearling weight in sheep (h^2 0.35) with a genetic correlation between them of 0.17. Suppose also that 10% of males and 40% of females were selected

Table 7.5 **Indirect selection for litter size in sheep (example)**

| | Selection differentials for | |
Feature	Litter size	Yearling weight
Females	0.966	0.966
Males	0	1.755
Average	0.483	1.360

then you could assess the direct results of selecting for litter size and the indirect results upon litter size with selection for increased yearling weight.

Selection differentials in standard deviations would be as shown in Table 7.5.

The relationship between direct selection for litter size and indirect selection (by seeking increased yearling weight) would be:

$$\frac{CR_A}{R_A} = 0.17 \times \frac{1.360}{0.483} \times \sqrt{\frac{0.35}{0.10}} = 0.90$$

In this case correlated response would be about 90% of that achieved by direct selection largely because the genetic correlation between the two traits is low in this species. Nevertheless, selection pressure can be much greater in yearling weight because males and females can be assessed whereas in litter size only females can be measured. In this instance correlated response has not outperformed direct selection but in instances where selection pressure or heritabilities differ more widely that can occur.

Tandem selection

The simplest way of selecting for many features is to select each in turn. Thus trait A is selected for some generations and then trait B is selected. After some further generations trait C is selected and so on. If the characters involved are unrelated then as A improves, B and C will stay at their original level. If there are positive relationships between the characters then as A improves B and C also improve, albeit less so than A. However if traits are negatively related then as A improves B and C become worse. Selecting to correct the problem in B will lead to loss of the gains already made in A. Tandem selection is thus only really useful if few traits are involved and they are all positively related or, at worst, unrelated. Since many important traits are negatively related tandem selection has limited application.

Independent culling levels

This technique involves setting a minimum level of performance for each trait of interest. Any animal which fails to meet this minimum standard in *any* trait is automatically culled. The system has the advantage of being easy to operate even with large numbers of traits. It does, however, have the drawback that an animal which is outstanding in a particular trait but fails, even slightly, on another would be culled. The procedure using only two traits is illustrated in Fig. 7.2.

Individuals in segment 1 are used for breeding and all others discarded. In using this system it is important to limit traits to those of major interest and to set realistic targets. Little is gained by setting targets that few animals will achieve, especially for relatively minor features. The system does operate better than tandem selection when traits are negatively correlated.

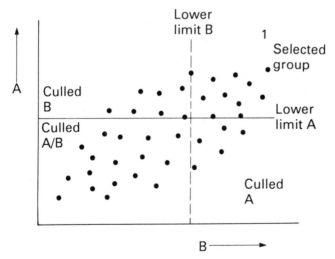

Fig. 7.2 **Independent culling levels illustrated for two traits**

Selection index

Perhaps the best technique for multiple objective selection is the selection index. Its disadvantage is that construction of selection indices is highly complex and certainly requires not only considerable data on each parameter but complex computer assistance in evaluation. Once constructed, selection indices are simple to use but their construction is not easy.

To construct an index you need the following data:

- Heritability of each trait in the index.
- Phenotypic variance of each trait.
- Genetic and phenotypic correlations between all traits.
- Relative economic value for each trait (this is not essential unless economic merit is required).

In any selection programme, genetic parameters may not exist at the start, and established (and perhaps unreliable) figures from other populations may have to be used until data are accumulated for the population under study. Difficulties also exist over relative economic value. This is based not on actual prices but relative prices; how much is a unit of gain worth compared with a unit of feed efficiency or a unit of lean meat production? Price changes of feedstuffs or products will alter relative economic merit and thus lead to a need to revise index data.

The aim of an index is to give the best prediction of an animal's breeding value. Correctly made an index maximises the correlation between the true breeding value and the prediction of the breeding value which can be termed the accuracy of evaluation. Essentially an index is an exercise in weighting the relevant items into a single figure. The Meat and Livestock Commission (MLC) has developed an index for selecting beef bulls which takes into account the ease of calving, birth weight, 200- and 400-day weights, feed intake, muscling score and conformation.

Genetic theory suggests that an index is more efficient than independent culling levels and this efficiency increases as more traits are involved. If n is the number of traits then an index is $\sqrt{n}$ times as efficient as independent culling levels. With only two traits an index would be 1.41 times as effective, but with 16 traits this would increase to 4.0 times.

A problem with selection indices is that they can only be applied to a specific set of circumstances and the compilation of data from very different populations is not feasible. Data collected in a test station for example will not be applicable to on-farm testing for which different indices would be needed.

Specialised lines

A fourth way of selecting for multiple objectives is to develop specialised lines. These are usually male and female lines, selected for different objectives, which are then crossed to give the commercial generation. Usually such lines are intended for meat purposes. By developing male and female lines fewer traits are selected in each which makes for easier and more efficient selection.

In general male lines would be selected for features such as growth, feed efficiency and carcass traits while female lines would be selected for reproductive traits and maternal aspects. Crossing the two should take advantage of heterosis (see Chapter 8) and bring about progress in all characteristics.

In some respects specific breeds could be regarded as specialised lines. In beef cattle, for example, Charolais are usually used as terminal sires rather than as the female side where smaller crossbred cattle might be wanted. Similarly, in sheep Suffolks correspond to the Charolais while in pigs female lines in Britain are mainly of Large White/Landrace origin while male lines may contain such breeds as Pietrain or Hampshire, useful for meat aspects but less successful in reproductive terms for a female line.

Best Linear Unbiased Prediction (BLUP)

The difficulty with selection indices is in trying to compile data from different sources such as different herds, breeds, sires, etc. If data were corrected for source all would be well but in many cases it is not known what correction factors should be used. Comparing animals born in 1988 and 1989 requires correction for year of birth and these data are specific to these years. The same is true of season of year or herd/flock of origin. With numerous years, herds, seasons to consider and widely differing numbers in each, merely assessing correction factors becomes a major statistical exercise. All of this needs to be done even before sires/dams can be assessed as to their breeding value and then ranked in order.

The Best Linear Unbiased Prediction or BLUP as it is known was developed by C.R. Henderson (1949, 1973) in the USA. In effect it has all the attributes of a selection index but it is much more powerful. As long as there are some common sires (called reference sires) used in the differing environments (herds/years/seasons/diets, etc.) then estimates can be made for a variety of issues. BLUP allows assessment of herd effects, year effects, genetic trends, biases caused by culling or selection as well as evaluating sires which were not necessarily used in the same locations. Increasingly BLUP has become the method of choice for assessing the Estimated Breeding Values (EBV) of sires in the cattle and pig world and now in the sheep industry. It is beyond the scope of this book to look at BLUP techniques but useful explanations are given by Nicholas (1987).

Selecting for threshold traits

When threshold traits are being considered, selection is made difficult because the phenotypic appearance gives no clues to underlying genetics except in the case of individuals showing the threshold trait. If one is selecting against a threshold trait which appears in, say, 10% of the population then selecting 15% of the 'normal' (unaffected) animals will not be equivalent to breeding from the 'best' 15% because you are only selecting 15 of the best 90% and therefore the 15% retained will be a random sample of the 90% 'normals'.

Selection in these cases is difficult and family selection may be more advantageous than individual selection. Taking into account family/pedigree data may allow selection of those individuals which do not appear to be closely related to affected animals or which come from families in which the incidence of the threshold trait is much lower than average. This may be a rather empirical way of selection and is far from ideal but there are no real alternatives. Threshold defects that are considered important need to be recorded in order that family incidence can be estimated and selection against the defect more accurately undertaken. Selecting animals unrelated or less closely related to affected stock or from families with a low incidence will result in a gradual decline in the threshold trait assuming its heritability is moderate to high.

Genotype–environment interaction

One of the many problems met with in animal breeding is that of where and in what conditions to test animals? For example, will performance testing of bulls on high concentrate diets be useful for identifying those bulls whose progeny will do well on pasture? Will pigs doing best when fed *ad libitum* diets be the pigs which will do best on restricted diets? A more extreme example would be whether dairy cattle selected under temperate conditions will be more useful in tropical conditions than dairy cattle actually selected in the tropics.

All of these questions and many similar ones are concerned with genotype–environment interactions. Confusion arises if breeders believe that effects upon the animal phenotype also influence the genotype. Cattle fed on pasture are unlikely to grow as quickly as those fed on high cereal diets but this is a phenotypic effect. The genotypes of the cattle are unchanged in the same way that artificially removing horns will not lead to their absence in the next generation.

A genotype–environment interaction exists when the ranking order of individuals or breeds changes in different environments. If, on pasture, the ranking order of seven bulls (or breeds) was 1, 2, 4, 5, 6, 3 and 7, but on concentrate diets the ranking order was 7, 3, 6, 5, 2, 4, 1 there would have been an almost complete reversal and it would be unwise to select other than in the environment in which the bulls (breeds) were to be used.

Another example is 'cancer eye' (bovine squamous carcinoma) in Herefords which is a genetic trait but one in which bright sunlight may have a bearing on the expression of the disease. Hereford breeders in Britain have no cause to worry about this defect since it will not appear in the British climate. However, Herefords in Texas which carried the right genotype express the feature. Accordingly selection of Herefords in Britain for export to Texas would give no clues as far as 'cancer eye' was concerned. If 'cancer eye' genes were present in the British animals this could not be selected out in Britain because the disease cannot be expressed.

Because there are so many environments and genotypes it is impossible to say that genotype–environment interactions will not occur in a species. Experimental evidence suggests that dairy and beef cattle have few important genotype–environment interactions whereas these are more important in pigs and sheep.

If genotype–environment interactions are important then it is usually better to select in the environment where stock will be used. If stock are used in many widely differing environments, climates or management regimes there is a great responsibility upon breeders to be aware of the possibilities of these interactions.

8 Breeding Systems

Introduction

Having decided which animals to breed from, breeders must decide how these will be mated. In broad terms breeders can decide whether to mate animals of the same breed or different breeds and if choosing the first policy can decide whether to mate related or unrelated individuals. The possibilities are summarised in Table 8.1.

Species crossing

This procedure has rarely been exploited in livestock breeding. There are difficulties in using species with different chromosome numbers since embryo survival is usually low and even if there is survival there is often sterility. Some crosses of zoological interest have been undertaken such as lion × tiger (liger or tigon) and wolves × domestic dog, but few crosses of farm livestock species have been undertaken.

The most famous is the horse × donkey giving rise to the mule. There have been crosses of cattle and buffalo, largely undertaken in the USA and Canada, called beefalo or cattalo. Crossing of zebu and the yak has been undertaken in the Himalayas and some crossing of sheep and goats has been made with the aid of genetic engineering. In the main these crosses are of minimal interest and are comparative rarities.

Breed structure

Before discussing cross breeding or purebreeding, some explanation of breed structure is needed. Breeds are of relatively recent origin

Table 8.1 **Breeding systems that can be considered**

Mating different species	Mating different breeds	Mating the same breed
Species crossing	Crossbreeding	Mating relatives
	Grading up	inbreeding
	Backcrossing	linebreeding
	Crisscrossing	Mating unrelated stock
	Rotational	outbreeding
	Lauprecht	mating likes
	Gene pool	mating unlikes

with the Shorthorn herd book, started in 1822, being the first such publication. Pedigree breeds as such can only legitimately exist as far back as pedigrees go.

The definition of what constitutes a breed is variable but generally refers to a group of animals within a species which have certain characteristics in common which render individuals of that breed recognisable as such. Most of these characteristics are physical ones connected with coat colours and the like such that at the post-slaughter stage with the hide removed breed identification is much less easy. Many breeds have arisen because physical isolation has left the population separated from other populations and hence selection or random drift has caused distinctions to appear. Other breeds have evolved by crossing among older established breeds until a new type has emerged. The dog is possibly the best example of a species which has been subdivided into numerous remarkably different breeds. The Chihuahua and the St Bernard belong to the same species and could be mated (albeit with some practical difficulty) to provide viable offspring. No other .species could claim such physical extremes. Although canine pedigrees could only trace back, at best, to 1873 (when the British Kennel Club was formed) some breeds of dog have existed in something like their present form for centuries but without pedigree proof.

Once a breed was established, breed societies were usually set up. These are organisations operating to foster the interests of a particular breed. Evidence suggests that all breeds in all species have developed a triangular-like structure usually termed a breeding

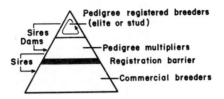

Fig. 8.1 **Traditional breed structure**

pyramid. At the apex are a few 'elite' herds/flocks which have tended to be either self-contained or which interchange stock with each other. These elite groups, sometimes called *studs* or *nucleus herds* generally sold stock to a larger group below them termed *multipliers*. These, in turn, sold stock to the commercial producers which may have been purebred but which were usually not registered. Generally movement was downwards only and the pedigree registration barrier between multipliers and commercial producers prevented upward flow at this level. Sometimes herd/flock books were 'open' which meant that breeders with other breeds could, by a series of topcrossings (usually four or five) with sires from the breed aspired to, enter upgraded stock in the pedigree society. The pattern of breed structure is shown in Fig. 8.1.

When AI was first established in cattle it was felt that using AI sires originating in the elite group would pass genetic merit more quickly to the commercial breeders. However it was usually found that the so-called elite units were not necessarily genetically superior to others in the breeding pyramid. Early investigations into the British Friesian breed, for example, showed that the elite group were essentially herds which had obtained stock from the 1936 importation from Holland, regardless of the merit of that stock. After some years of AI use, no improvement occurred and the nature of breeding pyramids in pedigree breeds was exposed.

The problem with the breed structure as shown in Fig. 8.1 is that in closed herd/flock books there is no upward flow of new or better genes. While elite units may have had excellent phenotypic performance they were not generally very superior, if at all, in genetic terms and frequently elite units were too small to make much genetic progress. The registration barrier was an artificial one based upon 'pedigree' rather than performance.

This said, the pyramidal structure has much to recommend it.

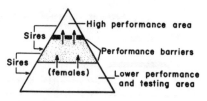

Fig. 8.2   **An improved breed structure**

Testing for genetic merit is costly and may not be feasible on a large scale. If testing was confined to an elite group selected on performance then this merit could filter down through the pyramid towards the commercial units. This is illustrated in Fig. 8.2.

Breed societies are part of the agricultural tradition in many countries. Generally they have taken an administrative and moral responsibility for the breed in question. They have acted to keep records of pedigrees and thus ensure breed purity, and they have served as a focus for breed promotion through a variety of activities as well as a kind of social club for those interested in the breed.

On the debit side, pedigrees without performance data are of minimal value and the aspect of self-preservation inherent in breed societies has tended to make them conservative about change and often impervious to ideas with breed councils (ruling committees) generally made up of older-established breeders unwilling to alter their views. In some cases AI was opposed, as it still is in most horse societies, and such things as contemporary comparisons were only brought in against the wishes of some breed societies.

In recent years the concept of pedigree and breed has disappeared from the poultry industry, and is quickly being undermined in the pig industry. In countries like Britain, the advent of new beef breeds from Europe, starting in 1960, served as a spur to breed societies to become more progressive and there has been a quickening of the acceptance of new ideas. Livestock shows, at one time based solely on physical appearance and often exaggerated appearance at that, have been changing, albeit slowly. Shows provide a meeting place for breeders with common interests and an element of competition that should spur some to better things. Increasingly, traits other than type are considered, though the confounding influences of herd/flock of origin still cannot easily be overcome. Many societies now openly collaborate on scientific evaluation in a variety of ways and this has not only breathed new life into previously moribund organisations but has actually led to greater understanding of animal breeding principles and performance advances.

Purebreeding systems (inbreeding)

Purebreed or pedigree breeders are, by definition, going to use individuals of the same breed. They can either use related (*inbred/linebred*) or unrelated (*outbred*) animals for that purpose.

In the establishment of breeds inbreeding was frequently practised at high levels. Inbreeding is defined as the mating of individuals more closely related than the average of the population from which they come.

That individuals of the same breed are related in some way is obvious from the nature of a pedigree. With two parents, four grandparents, etc., there are, by generation 20, over a million ancestors in that generation alone and over two million in the next. In a cattle breed 20 generations could be about 100 years and that long ago there were probably not a million members of the breed. Clearly many ancestors must be duplicated and hence relatives have been mated together. For this reason the definition refers to 'more closely related than the average'.

Inbreeding is measured by the inbreeding coefficient devised by Sewell Wright in the 1920s. An illustration of how to calculate this is given in Appendix 1.

As a system, inbreeding arouses a great deal of anxiety among lay persons and some practical breeders. In humans there are limits to the closeness of mating with first cousins being the closest allowed (inbreeding coefficient 6.25%) but there are no limits in animal breeding. However, certain problems can arise with inbreeding.

The first of these concerns deleterious genes causing defects. Inbreeding does not actually cause defects but because many defects tend to be recessive in nature, the mating of closely related stock increases the chances of such deleterious genes coming together and thus giving rise to abnormal progeny. Since all individuals will carry a number of deleterious genes, the occurrence of abnormalities is inevitable with inbreeding, but that in itself is not a major disaster.

Defects have to be assessed in relation to their seriousness and the qualities of the animal which produced them. If a dairy bull is an outstanding improver of milkfat/protein yield then it would be ludicrous to cull him, simply because once in every 10 000 progeny he produced a defect like 'amputate' which is a dead calf born with amputated feet and a parrot-like muzzle. The loss of calves, even if this occurs more frequently than one in 10 000, is of minimal economic importance in relation to the benefits the bull can give. Those who become emotionally involved on such issues must realise that all stock carry defects and there is no way they can all be eliminated without ceasing all breeding.

Of greater importance than the occasional defective progeny is the consequence of inbreeding depression. This is the gradual lowering of performance with increasing inbreeding. Usually the traits affected are those concerned with 'fitness' which tend to be traits of low heritability but highly influenced by non-additive features (especially dominance). If inbreeding is undertaken then such things as fertility could decline, embryonic mortality could increase, progeny survival decline, growth rate be lessened, milk yield decline whereas carcass traits might be unaffected.

Different species vary in the effect inbreeding has but it is likely that pigs will show problems at lower levels of inbreeding than would

Table 8.2 **Effects of inbreeding in sheep**

Trait	Number observations	Regression coefficient	
		individual	*dam*
Greasy fleece weight	>6356	−0.017 kg	0.005 kg
Clean fleece weight	>8054	−0.001 kg	
Staple length	>13612	−0.008 cm	0.002 cm
Birth weight	3678	−0.013 kg	−0.013 kg
Weaning weight	10183	−0.111 kg	−0.072 kg
Post-weaning weight	>5777	−0.178 kg	0.013 kg
Body type score	9902	0.011	0.010
Condition score	9902	0.012	0.017
Face cover score	>11902	0.002	0.014
Wrinkle score	>12630	−0.020	0.002
Lamb survival	>6266	−0.028	

Regression shows change in performance for each 1% increase in inbreeding of the lamb or of the dam.
Source: Lamberson & Thomas (1984).

cattle. Individual herds/flocks vary so you cannot be dogmatic on this score, but in any species inbreeding poses problems for individual breeders and the higher the level of inbreeding the greater the risk. You would expect few problems up to about 10% inbreeding and increasing difficulties in excess of 20%, but there are theories that it is not just the absolute level of inbreeding but the speed with which it is achieved. A level of 10% achieved inside three or four years may be more serious than 10% in ten years.

Lamberson & Thomas (1984) published a review of inbreeding effects in sheep based on 25 studies covering over 25 000 animals. A summary is shown in Table 8.2.

In most of the traits inbreeding had an adverse effect except on wrinkle score with greatest harm expressed on lamb survival.

Linebreeding seems to arouse fewer fears and it is often said that if mating relatives fails then blame it on inbreeding, but if it succeeds credit it to linebreeding. In reality inbreeding and linebreeding are the same thing, differing solely in degree. If inbreeding is dangerous, linebreeding will be less so. If inbreeding succeeds, as it sometimes can, then linebreeding will succeed more slowly.

The levels of inbreeding which will occur in herds/flocks that are closed to outside material and use varying policies are shown in Table 8.3.

Table 8.3 illustrates how very close matings (e.g. full or half-sib matings) lead to a rapid build-up of inbreeding whereas, even with relatively few sires used annually, inbreeding does not build-up rapidly. For the 1-, 3- and 5-sire herds/flocks a generation length of

Table 8.3 **Intensity of inbreeding (%) for different closed herd/flock programmes**

Generation	Offspring/ parent or full sibs	Half-sibs	1-sire 3-sire 5-sire herds or flocks new sires per year (five year generation)		
1	25.0	12.5	2.5	0.8	0.5
2	37.5	21.9	5.0	1.6	1.0
3	50.0	30.5	7.5	2.4	1.5
4	59.4	38.1	10.0	3.2	2.0
5	67.2	44.9	12.5	4.0	2.5

five years is assumed. Thus 1, 3 or 5 new sires per year means 5, 15 or 25 new sires respectively per generation. It is also assumed that mating is random in these units, but by careful attempts to reduce close matings the build-up would be less than that given.

Inbreeding increases prepotency which is defined as the ability of an individual to stamp its virtues (or defects) on its progeny. It is possible for inbreeding to be a powerful tool for advancement if used as part of an eventual crossing programme using inbred lines. Inbreeding was also much used in the establishment of particular breeds, but it is not a breeding system to be used by the uninitiated and unwary since it can go disastrously wrong.

Inbreeding levels are much less in most breeds than many people imagine. You can divide inbreeding into current (i.e. within the first two generations and therefore undertaken by deliberate policy) and non-current inbreeding. This latter can be divided into long-term inbreeding and inbreeding due to the separation of the breed into strains. Even if breeders make no attempt to inbreed they will be using individuals that show some relationship because the breed is effectively a closed unit (albeit large). Studies undertaken in a variety of species and breeds suggest that the increase in inbreeding per generation for many, lies somewhere between 0.25 and 0.7% with most around 0.35%.

Unfortunately many schemes set up to advance breeds cannot always be large enough in terms of animal numbers to always avoid inbreeding, and it would be feasible for inbreeding depression to be large enough to offset selectional advances.

Purebreeding systems (others)

Outbreeding is the mating of animals of the same breed but with minimal relationships between them. It is frequently undertaken but if it runs into problems you cannot always see where to go next, whereas if inbreeding runs into problems in one generation a judicious mating can bring the inbreeding coefficient back to zero. Calculation of the coefficient of relationship is shown in Appendix 2.

Mating of like to like is termed assortative mating. It is undertaken where breeders seek to mate animals with the same virtues. It is often called mating the best to the best. Strictly the system not only involves mating the best to the best, but also the worst to the worst. Usually the worst are discarded from a breeding programme so that assortative mating involves some selection. Animals that look or perform alike may not carry the same genes, and hence this system does little to alter gene frequency.

Mating of unlike to unlike is often called compensatory mating. Most breeders use it at some stage and seek to correct failings in an individual by mating it to one excelling where the first fails. This policy tends to make the population more alike but, as with the mating of likes, it has limited influence in altering gene frequencies.

CROSSING BREEDS

General

The basis of crossbreeding is the exploitation of hybrid vigour or heterosis, the latter term being preferable. If two breeds or lines are crossed then you might expect the progeny to be at the midpoint between the two parental means for any given trait. If the progeny exceed the mean of the two parents then the extent to which they exceed that midpoint is a measure of heterosis. In plants, heterosis is usually confined to progeny which exceed the best parent but this is rare in animal breeding. The crossing of a numerically small breed with a numerically large but inferior breed may give progeny that exceed the midpoint and the fact that they do not excel over the better parent does not mitigate against their value, since in one generation the numerically large but inferior breed is improved much further than might have occurred by direct selection.

Just as traits are depressed by inbreeding, so traits respond to crossbreeding by exhibiting heterosis. In general, characters that are non-additive will show the greatest heterosis and thus heterosis and inbreeding depression might be considered opposite sides of the same coin. Some estimates of heterosis are shown in Table 8.4.

Greatest heterosis is likely when the breeds being crossed differ markedly in their gene frequencies and the trait(s) being considered are under the control of dominance. Heterotic effects are often high when crossing *Bos taurus* and *Bos indicus* cattle breeds.

Table 8.4 **Some general estimates (%) of heterosis**

Species	Trait	Heterosis (%)
Dairy cattle	Milk yield	2–10
	Milkfat yield	3–15
	Feed efficiency	3–8
Beef cattle	Calving rate	7–20
	Calf viability	3–10
	Calves weaned/cow mated	10–25
	Birth weight	2–10
	Weaning weight	5–15
	Feedlot gain	4–10
	Carcass traits	0–5
Sheep	Barrenness	18
	Lambs born/ewe lambing	19–20
	Lambs weaned/ewe mated	60
	Birth weight	6
	Preweaning growth	5–7
	Carcass weight	10
	Fleece weight	10
Pigs	Piglets/sow farrowing	2–5
	Litter size/weaning	5–8
	Litter wt/weaning	10–12
	Growth to slaughter	10
	Carcass traits	0–5

After Dalton (1985).

Grading up/topcrossing

These two techniques are very similar. Topcrossing usually refers to a breeding system in which the breeder returns to the 'original' line to gain new genetic material. Friesian breeders in Britain returning to Holland for bulls would be topcrossing as would American Hereford breeders returning to England for bulls.

Grading up is the gradual changing of a breed by continued crossing with another. This system has been used all over the world to 'upgrade' one breed and convert it to another. Most breed societies with open herd/flock books would accept four or five topcrosses of this kind to consider the upgraded animal purebred. After four crosses the animal would be 93.8% pure. The better the sires used in such a programme, the better the end product, but if the new breed was considerably superior to the existing one even the use of random sires would result in progress.

It is established that the major effect of heterosis is in the F1 generation, i.e. the original crossbred. Crossing F1 with F1 (to give F2) results in some further advantage due to the mother now being crossbred as opposed to purebred, but heterosis in the individual F2 is halved. Accordingly, heterotic effects in grading up are eventually lost since the net effect is to change from one purebreed to another.

Backcrossing/crisscrossing

A backcross occurs when a first cross animal is mated back to one or other of its parent breeds. This has been widely practised with pigs in Britain where Large White × Landrace females were mated to either Large White or Landrace boars. Although the resultant slaughter generation is 75% of one breed and 25% of the other, it is a scheme intended to take advantage of using a crossbred dam.

Crisscrossing is an extension of backcrossing. Initially two breeds (A and B) are crossed. The crossbred is then mated back to A to give 75% A 25% B. This crossbred is then mated back to B to give 62.5% B and 37.5% A. Thereafter breeds A and B are used in alternate generations. The policy was once quite widely used in American pig breeding and many new breeds were established by crossing techniques coupled with selection.

Rotational crossing

In this system a series of breeds (three, four or more) are used in succession. This is shown diagramatically in Fig. 8.3.

In principle, by using several breeds there ought to be greater heterosis than by using a two-breed backcrossing or crisscrossing policy. Experimental evidence suggests that rotational crossing with three or four breeds has proved somewhat superior to two-breed

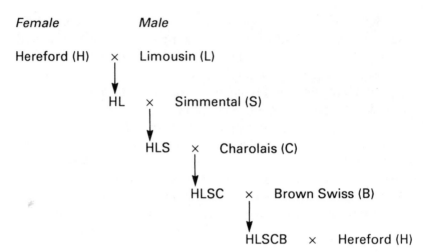

Fig. 8.3 **Rotation crossing using five breeds**

crossing. However, much will depend upon the breeds available in any species and their compatibility.

The Lauprecht system

This system was put forward by a German scientist (Lauprecht, 1961) and essentially consists of a crossing programme using three breeds (A, B and C). A is crossed with B and the resultant crossbred

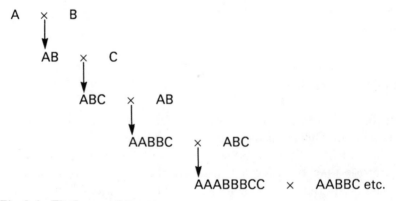

Fig. 8.4 **The Lauprecht system**

mated to C. Thereafter the policy is slightly different to other programmes because only crossbred males are used and in each case the female generation is mated to males from the previous crossbred generation (see Fig. 8.4).

Essentially this system results in a population which, by about the fourth cross and all crosses thereafter, is approximately one third of each breed. Although this is an unusual programme, the disadvantage is that it effectively becomes a new breed rather than a distinct crossbred. Whether it has been used successfully in a practical context is uncertain.

Gene pool

This system developed in Edinburgh with a pig population. The particular trial started with four distinct breeds and selection was practised for certain features including reduced backfat thickness. Animals were selected solely on merit rather than on breed or cross so that crossbred animals could be used. New breeds were also added at a later stage. There is much to recommend such a policy since it is based solely on performance as the criterion for use. It is important to appreciate that new breeds need to be incorporated solely on merit and not simply because they are new.

Creating new breeds

Although most breeds in a species trace back to the 19th century or earlier, new breeds are still being formed. Sometimes breeds are little more than a cross between two foundation breeds with interbreeding among the F1 animals. The New Zealand Half-Bred is a case in point being a cross between the English Leicester and the Merino. The expected decline in the F2 needs to be offset by selection within the population.

Other new breeds have been developed for specific situations which is especially true in cattle where attempts to find a combination of *Bos indicus* and *Bos taurus* suitable for the tropics has been much practised. Often there has been a deliberate attempt to use about $\frac{5}{8}$ *Bos taurus* and $\frac{3}{8}$ *Bos indicus* though there is nothing 'magical'

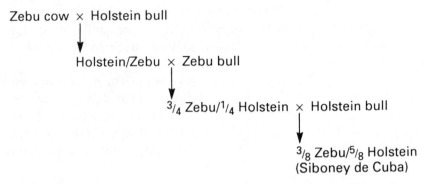

Zebu cow × Holstein bull

Holstein/Zebu × Zebu bull

$^3/_4$ Zebu/$^1/_4$ Holstein × Holstein bull

$^3/_8$ Zebu/$^5/_8$ Holstein
(Siboney de Cuba)

Fig. 8.5 **Method of producing the tropical milking breed – the Siboney de Cuba**

about this particular combination. Examples (not all $\frac{5}{8}$; $\frac{3}{8}$) with area of origin are:

- Santa Gertrudis cattle (Shorthorn/Brahman) – Texas.
- Luing cattle (Shorthorn/Highland) – Scotland.
- Australian Milking Zebu cattle (AMZ) (Jersey/Zebu) – Australia.
- Jamaica Hope dairy cattle (Jersey/Sahiwal/Holstein) – Jamaica.
- Colbred sheep (East Friesland/Clun Forest/Border Leicester/ Dorset Horn) – England.
- Brangus cattle (Angus/Brahman) – USA.
- Charbray cattle (Charolais/Brahman) – USA.
- Siboney de Cuba dairy cattle (Holstein/Zebu) – Cuba.

The way in which the Siboney was produced is illustrated in Fig. 8.5.

Exactly how successful all these 'new breeds' are is open to question. The idea of producing a new breed for specific environmental conditions may seem highly attractive but such programmes take 20 years and more. By the time that a breed has been developed for 'today's' conditions it is actually 'tomorrow', and those conditions may have been changed by better nutrition and management. Frequently new breeds are gimmicky and greater progress might have resulted from the same investment into management improvements.

9 Breeding in Practical Terms

General principles

Because each breeder is faced with different circumstances in terms of species, herd/flock size, location, management techniques and objectives, it is not possible to lay down hard and fast rules for any given set of circumstances. The purpose of this chapter is therefore to look at broad overviews.

You have to distinguish between breeding objectives and selection criteria. In drawing up an objective you have to look at the enterprise as a whole taking account of the production system, the inputs to it and the outputs from it including all forms of income derivation. The decisions about what traits to select for ought to be determined by their relevance to the overall economic objectives, rather than because they are highly heritable or easy to measure. Ponzoni and Newman (1989) examined this in relation to beef cattle but the principles have much wider application. Those traits in the breeding objective may be considered the ends, while the characters used in the selection criteria are the means of achieving those ends.

It has been shown that progress depends upon the heritability of the character, the selection differential and the generation interval, all of which, to a greater or lesser degree, can be influenced by the breeder. It has also been shown that routes to progress are:

(1) Selection of males to breed males.
(2) Selection of females to breed males.
(3) Selection of males to breed females.
(4) Selection of females to breed females.

In general terms route (1) is the one with greatest opportunity for progress closely followed by (2). Route (4) is the least useful unless embryo transfer is fully available.

Most breeders are in control of the female side of things since they will have a herd/flock, and they will decide which females are retained for breeding and how they will be used. Males may also be bred by the breeder but may also be purchased from other breeders or companies; while in several species males may be used in the form of semen purchased from private organisations or government bodies. In this case, while the breeder still has to make the decisions about which semen to buy, the breeding programmes producing such sires may be in the hands of the organisation selling the semen. Sometimes the person actually farming the animals may not be a breeder in any strict sense, but simply buy in females/males of a specific line or breed which would be culled at the end of their breeding life to be replaced by other purchased stock. Increasingly this happens in pigs and it has been the norm in poultry for some years. The company/organisation producing the breeding stock will be where major genetic decisions are made.

Identification

In any programme accurate identification of animals is important and this cannot be overemphasised. Lost identification or misread tags can cause havoc with breeding programmes and result in the loss of valuable and expensive data.

Permanent identification is not always as permanent as it sounds but would include:

(1) metal ear tags,
(2) plastic ear tags,
(3) ear tattoos,
(4) freeze brands,
(5) fire brands,
(6) plastic tail, hock, ankle tags,
(7) ear-notching.

Not all these systems apply to all species. Ear notching, for example, is usually used in pigs and to a lesser degree sheep but can run into 'welfare problems'. Ear tags are very useful but can be lost, particularly plastic ones, though these have the advantage of being available in numerous colours. Tags are not very readable from a distance and metal tags require the animal to be restrained. Ear tattoos also require restraint and can become illegible as well as being dependent upon the ear colour. Fire branding is largely a beef ranching system which, though fairly effective, leads to damage of the hide. The same is true of freeze branding but brands show up as white hair and are best on dark cattle. Dairy cows are often freeze branded on the rear for ease of reading in the milking parlour. Plastic tags on tail, hock or ankle are usually dairy cattle features, also chosen for ease of reading during milking.

Semi-permanent identification systems include:

(1) Neck tags of wood/plastic attached by a cord.
(2) Hair dyes or bleach used on the animal's side.
(3) Paint brands.
(4) Plastic covered wire twisted into a hole in the ear.

Hair dyes or bleach numbers on the side (usually sheep) can be read at greater distances than most other systems, as can paints, but these should be used sparingly and only approved washable paints used. By definition semi-permanent systems are intended for relatively short periods of time.

Temporary markings would include stick-on labels and washable paint sprays or tie-on ribbons and chalk raddle marks. They are useful only during a very short period.

It is useful to have more than one system in operation. For example metal tags are frequently used in calves but they are not easily read. However a plastic tag can be inserted (various colours to indicate sire, group, dam age, etc.) and if the plastic tag is lost the metal tag is still there to prevent loss of data. The more animals that are in the programme and the more extensive the management system, the greater the risk of lost identification and hence inaccurate or lost information.

It must be realised that all of the above identification systems are used in live animals but are lost once that animal is slaughtered. Increasingly breeders require feedback on carcass traits and the lack of identification at the carcass stage is a barrier to progress. At present electronic implants are being examined in several species. Although more expensive to use and even more expensive to read, since the instruments required to read electronic implants are not cheap, this is the one way that could allow carcass details to be related back to the live animal. It is necessary to locate an implant where there will be no migration of the implant and it can be effectively read and eventually easily removed.

Dairy cattle breeding

GENERAL

Dairy cattle breeding is usually directed towards milk yield, milk quality and regularity of calving. All developed countries and many developing ones will have official milk recording systems which should be consulted. The more frequently milk is measured the greater the accuracy but weekly or monthly recording might be accurate enough for many purposes. Testing for quality requires laboratory facilities, but machines exist which can evaluate numerous samples very rapidly.

Milk yield and milkfat/protein yield are of moderate heritability while milkfat/protein percentage tend to be high. Regularity of calving is of low heritability. The fact that milk is assessed only in females complicates selection procedures and generally requires bulls to be progeny tested. Correlations between milk/quality traits tend to positive and high when in weight terms but milk yield and fat percentage are slightly negatively related.

Most dairy farmers would undertake some selection for traits such as temperament and cull those cows which are excessively nervous or which are slow milkers. Usually such animals will not yield well in any event because they will rarely be given the excessive time they require for complete let down.

Culling also takes place on the grounds of disease (e.g. susceptibility to mastitis) as well as on poor reproductive performance. In both of these cases there may be some association with yield as high yielders may be more prone to mastitis and reproductive problems. This is unfortunate but no herd has a place for cows of poor fertility or high disease risk. It does, of course, need to be remembered that failure to rebreed may, in some instances be a failure of stockmen to accurately identify cattle in oestrus.

It is useful to record heat periods, services and calving dates to check on fertility and in addition some record of calving difficulty is useful. This will have to be subjective on a scoring system e.g.:

0 = no assistance
1 = slight assistance
2 = veterinary assistance
3 = considerable assistance using mechanical/veterinary aid
4 = caesarian section

Birth weights of calves might be recorded along with gestation period, if known, as well as sex. Milk would be recorded by weight or volume and composition in absolute and percentage terms. The lactation length is also needed as well as the number of times milked per day. A three-times-daily milking might be expected to increase yield by some 10%. You cannot compare thrice- and twice-daily milking unless records are corrected.

Some aspects of type would be of value if these can be related to practical features such as resistance to disease or to longevity. Unfortunately longevity is not easy to measure and is likely to be of low heritability as well as having the practical problem of it taking a long time to actually assess.

Generally the best cows would be bred to bulls of the same breed and the poorer cows to beef bulls if there is a need for beef. Culling on yield would be undertaken at the end of each lactation but only after culling on fertility, disease, etc., had been done.

CONTEMPORARY COMPARISONS AND PROGENY TESTING

In the main, dairy breeding has been based upon progeny testing of sires using some kind of comparison between the bull under study and all other bulls whose daughters were milked in the same location at the same time. All progeny testing is based on comparing bulls in the same herd/year/season. In countries with smallish herds all records are used after correction to mature equivalent (the record expected had the cow been a mature animal). In other countries this is avoided by comparing only first lactation records. This latter reduces numbers but is possibly more accurate as first lactations tend to be unselected and thus relatively unbiased.

The British system is called the Improved Contemporary Comparison (ICC) and has been used in gradually altered form since the early 1950s. In this system heifers calving in the same herd, year and four month period (season) are considered contemporaries. Records

Table 9.1 Calculation of an ICC

Herd/year/Season (1)	Daughters of bull		Contemporaries				Calculation		
	Number (2)	Corrected yields (3)	Number (4)	Corrected yields (5)	ICC sire (6)	Corrected yield (7)	Difference (8)	Weight (9)	Weighted difference (10)
A	7	4500	3	4250	+240	4010			
			4	5000	+100	4900			
			7	4800	−50	4850			
				Average		4684	−184	4.67	−859.3
B	6	5100	2	4500	−100	4600			
			6	5200	+200	5000			
				Average		4900	+200	3.43	+682
C	3	5300	1	5200	+40	5106			
			4	5000	−100	5100			
			2	5400	+50	5360			
				Average		5175	+125	2.10	+262.5
Totals	16	4875						10.2	+85.2

of daughters of a bull (and all contemporary daughters of other bulls) are corrected for age at calving and for month of calving. A further correction is made for the ICC level of the contemporary sires and the data compiled into a single figure. Calculation is shown in Table 9.1.

The explanation is shown below for Table 9.1:

Column (1). Each letter represents a herd/year/season. A and B could be the same herd but different years or seasons.

Column (2) shows the number of first lactation daughters of the bull under test called a Limited Use (LU) bull.

Column (3) shows the mean corrected 305 day yields (kg) for the daughters. Correction is made for age and month of calving.

Column (4) gives the numbers of daughters of other bulls arranged in sire groups.

Column (5) is the average corrected yield for each sire group.

Column (6) is the ICC of the contemporary sire (kg).

Column (7) is the corrected contemporary yield after taking account of sire ICC. A + sire has his ICC deducted and a − sire has it added. The average for each herd/year grouping appears below. This is weighted by daughter group numbers.

Column (8) is the difference between the daughter average and the contemporary average.

Column (9) is the weighting or number of effective daughters of the LU bull. This is assessed using the formula:

$$W = \frac{\text{Number of daughters} \times \text{Number of contemporaries}}{\text{Number of daughters} + \text{Number of contemporaries}}$$

$$\text{For A this is } \frac{7 \times 14}{7 + 14} = \frac{98}{21} = 4.67$$

Columns (8) and (9) are multiplied to give the weighted difference Column (10). The apparent merit overall is then 85.2/10.2 = + 8.5 kg.

Although this bull appears to be raising milk by 8.5 kg in his daughters there is a need to take this back to a base year (now 1983) and to take account of the regression of future daughters upon present daughters. This will reduce the effect still further because the bull has relatively few daughters. This bull would be little better than average. Because the ICC works on daughter performance it is equivalent to half the bull's Breeding Value.

ICCs are now calculated for milk and milkfat and protein yield, as well as milkfat and protein percentage. Similar figures are calculated for some 16 type features using linear assessment evaluations. In these daughters are assessed for a series of features like stature, angularity, rear udder attachment, etc. Each animal is scored on a scale from 1 to 9 with these representing extremes. Therefore 1 would represent very small cows and 9 very large cows in terms of the stature feature.

Although other countries do not use ICCs most use similar techniques involving comparison of animals in the same herd/year/ seasons. Most developed dairy evaluation systems produce a great deal of information for breeders. Not only are sires listed with their apparent merit for a series of features but the ever increasing use of BLUP techniques (see Chapter 7) allows comparisons to be made between sires tested in different countries. Canadian, American, British and New Zealand tests can be readily assessed by AI organisations and data made available to breeders. This is illustrated by the MMB (1987).

COW GENETIC INDEX

The Milk Marketing Board of England and Wales (MMB) has developed a Cow Genetic Index (CGI). This was based on indices previously used by three organisations including the British Friesian Cattle Society. The index is based on the yield of milkfat and protein combined, and the higher the value the better the cow.

To assess an index for a cow her own records are used up to five lactations together with the latest ICC of her sire and the CGI of her dam. Additionally the average genetic level of the herd has to be calculated from the average merit of sires and dams used in the herd at the time of indexing. This is an attempt to allow selection across or between herds and not simply within herds. A herd index is assessed for cows and heifers separately.

Suppose in a herd the average merit for sires was + 6.7 kg (fat plus protein) while that for dams was −1.4 (fat plus protein) then the mean parental merit is (6.7 − 1.4)/2 = 2.6. This is multiplied by 10 to give a more manageable number and then 500 is added. The herd index would thus be 2.6 × 10 + 500 or 526. This indicates a slightly above average herd since an index of 500 theoretically represents an average herd.

Suppose we have a cow in this herd with a fat plus protein average level of 52, a dam at 0 and a sire at -22. This cow's genetic index would then be:

$$52 + 0 + (-22) + 526 = 556$$

MMB figures suggest that the mean CGI of Friesian cows is around the 570 mark with a standard deviation of around 96, though these figures alter annually. This would mean that cows exceeding $570 + 3 \times 96 = 858$ would be in the best 0.5% of the British breed. These might, of course, be outside the best group if other countries are considered such as North America.

The MMB annual reports of the Breeding and Production Organisation are full of data on the testing of dairy cattle and private firms such as the Dairy Herd Improvement Company also produce useful information. Similar publications exist in other countries and are invaluable to dairy cattle breeders.

MOET SCHEMES

The advent of multiple ovulation and embryo transfer has led to different ideas about testing dairy bulls. Nicholas (1979) and Nicholas and Smith (1983) published ideas for MOET schemes and these are now being set up in different locations. They are complex, but essentially they involve assessing potential sires on the basis of pedigree information together with lactation information on full and half sisters to the young bulls.

The main advantage of MOET schemes would be to reduce generation interval to about three years compared with about 6.5 in conventional progeny testing schemes. Progress depends upon the numbers of progeny used in transfers and methods of selection, but it was originally envisaged that progress from a scheme with about 1000 embryo transfers and 500 milk recorded females per year would give 30% greater progress compared with conventional schemes.

Essentially MOET schemes are an attempt to produce information from relatives along the lines of litter information available in species like the pig. MOET techniques are still being refined and tested and they require considerable skills as well as computer facilities. Disadvantages are the tendency to put the national scheme into a single unit and the disease risk that exists with cattle in one location. There are also considerable risks that, unless very large, these schemes could run into increasing problems from inbreeding. However, none of these disadvantages are beyond solving.

In Britain a private company Premier breeders set up a MOET scheme which was then taken over by the Milk Marketing Board. The scheme involves a 250 cow nucleus herd based on cows of North American origin with an initial CGI in excess of 1000 kg. From this unit 32 cows would be selected for embryo production using eight carefully selected sires. About 500 embryos would be produced from which some 130 daughters and 130 bulls would be available. The 130 daughters would enter the nucleus unit thus having a high turnover rate with no nucleus cow lasting beyond two lactations. Bulls would be selected on the basis of dam performance, the performance of four full sisters and 12 half sisters. Heifers would be selected on their own yield, dam yield, performance of three full sisters and 12 paternal half sisters. This would actually give more reliable tests on heifers than their brothers but the generation interval would be around the 2.5 to 3 year mark.

Revised estimates suggest that progress will be less than that envisaged by Nicholas and Smith (1983) but still on a par with progeny testing. Moreover, an open nucleus into which the best available cows nationally might be used could add progress over and above that from the nucleus unit itself. This open system would lower inbreeding over that of the closed nucleus unit. By starting with high CGI cows the scheme gets a boost and though it will take some years to be fully operational it could revolutionise dairy cattle breeding.

There are some political difficulties in that having convinced farmers of the value of progeny testing they are now being asked to use bulls on the basis of less accurate sibling tests. However the MMB will test two bulls of each family through the conventional dairy bull progeny testing scheme which will act as a measure of how superior MOET stock are. One advantage of MOET schemes is that in addition to the usual output data on milk/milk quality some input data on feed intake will be assessed and thus give extra benefits to the breeding programme. Although other criteria can be measured in nucleus cows the ability to use such items (fertility, longevity) even if measured is limited by being assessed on too few cows. Such features need to be assessed on the large scale of progeny testing schemes.

Beef cattle breeding

A carcass is the eventual end product of a beef enterprise but many beef producers may consider the weaned calf as their end product. The enterprise is by nature more extensive than dairying in that beef herds are not 'seen' on the daily basis that dairy cows are and such things as artificial insemination less readily applied.

Beef breeders are interested in reproductive performance, ease of calving and milking ability in the beef cow coupled with growth potential and carcass data in the progeny of those cows. Increasingly beef cattle tend to be crossbred since this is a means of improving reproductive performance which is of low heritability and thus unlikely to respond to direct selection. Calving difficulty is subjectively assessed (see this chapter, under *Dairy Cattle Breeding*) but milking performance is indirectly evaluated by looking at calf growth. Many authorities suggest small to medium-sized cows rather than large ones because the former are lower in maintenance requirements even though calves from them tend to be smaller. Dystocia is likely to be less in smaller rather than larger cows.

Calves need to be tagged, have their sex recorded and ideally a birth weight taken within 24 hours of birth. Although weighings can be taken at intervals it is usually only necessary to take a weaning weight together with an age (days) at weaning. In most situations weaning age is about the 200 days of age mark and weaning weight is usually corrected to that age by the simple correction:

$$\text{200-day weight} = \frac{[\text{weaning wt} - \text{birth wt} \times 200]}{\text{age (days) at weaning}} + \text{birth wt}$$

This corrected 200-day weight should be further corrected for dam age, usually correcting to a 5-year-old dam standard. Sex corrections can be also applied or sexes assessed separately.

Weight of calf weaned (WCW) is the usual selection criterion since it includes measures of production and cow fertility and as such it is an easy trait to measure with a moderate heritability. In reality it is a complex trait comprising a great many sub-traits but it is the main income of the beef herd.

Post-weaning performance can be assessed in bulls and heifers and various policies exist. In the USA bull testing tended to be based upon feedlot performance assessing daily gain over a period of about 148 or 196 days. In Britain assessment was based upon weight at 400 days of age. In Sweden and in Cuba assessment was based upon earlier weaning (90 or 100 days of age) and subsequent performance for either 200 days on test or for gain to a fixed weight (say 400 kg).

Correlations tend to be positive and high between 200- , 400- and 500-day weights and with birth weight such that increasing size at any age tends to be associated with an increase in size at other ages. Relationships are higher the closer the two age groups are to one another. Similarly, growth rate will be positively correlated to weight at given ages. The 400-day weight has the advantage of being close to puberty and is the general choice. In most countries simple selection for gain to a weight/age has been replaced in recent years by index selection, taking account of weights at different ages as well as feed intake and aspects of conformation. BLUP techniques are also being applied to beef cattle especially in Australia when a programme called Beefplan has been established.

At the carcass end, selection is usually directed towards reduced fatness/increased leaness either by taking assessments of fat depths (e.g. at the 10th or 13th rib) or by ultrasonic testing for eye muscle area. Increasingly ultrasonic testing of sires is being undertaken. It is extremely difficult to alter the distribution of muscles in the animal because within a species there seems to be only minimal differences between breeds. Progeny testing for carcass traits can be undertaken and also for dystocia although this latter usually requires a large number (about 300) of calvings.

Heifers are usually grown to a specific size before mating and selection is undertaken on the basis of weight gain. Those heifers which are too light for mating and would thus have to be mated later and to some extent out of the normal season of breeding, are best culled.

Once in the herd, beef cows are best selected on regularity of calving, ease of calving and on the quality of calf they produce which probably means weaning weight. Although regularity of calving is not highly inherited it is commercially undesirable to retain cows which breed irregularly.

Sheep breeding

In most countries sheep are bred as meat producers but all sheep breeders have some income derived from wool and in some countries (e.g. Australia), the primary role of sheep is as wool producers.

Wool sheep can be divided into those producing fine quality wool (the Merino being the best example), general purpose wool (many British breeds fall into this category) and carpet wool breeds (the New Zealand Drysdale is a specialist breed in this connection and many British hill breeds fit to some degree into this category).

General-purpose wool breeds are usually selected for a combination of numbers of lambs weaned (NLW), lamb weaning weight (LWW) and some aspect of carcass quality (leanness). Some attention is also paid to greasy fleece weight (GFW) but little towards the improvement of fleece quality. Although lamb numbers are not highly heritable, other traits would be medium to high in heritability terms. The weight of lamb weaned (WLW) might be used in place of numbers of lambs and individual weaning weights. It is moderately heritable and takes some account of fertility. In New Zealand breeds of this kind are often index selected on the basis of:

$$NLW + LWW + LDP + ECW + GFW$$

where LDP is lamb dressing percentage and ECW is ewe carcass weight.

Fine-wool breeds are probably best selected for similar traits as general-purpose wool breeds because income derives from both meat and wool. However, greater importance needs to be placed upon wool quality. Increasingly the Bradford count is becoming outdated and quality is assessed on fibre diameter (FD) and staple length (SL) since these are of prime importance in the manufacture of quality garments. A fibre diameter of 34 microns or finer is desirable and selection is best directed towards increased fleece weight but with an upper limit set to fibre diameter. An index might thus be based upon:

$$NLW + LWW + LDP + ECW + GFW + SL + FD$$

Carpet wool breeds have been developed in some regions. Such wool is heavily medullated, i.e. has hollow fibres (hair) rather than the solid fibres of wool. Breeding objectives would again be similar to general purpose wool breeds but with emphasis upon carpet wool qualities.

The problem with medullation is its assessment. Usually this is subjectively done by eye rather than by any objective technique and such methods are not always very accurate. Part of the problem stems from the failure of manufacturers to stipulate exactly what they require in carpet wools. These can be classified as heavily, lightly or variably medullated, but the exact specifications of medullation, bulk, staple length, lack of lustre and harshness need to be quantified more precisely than at present. Increasingly this is happening.

There is evidence that extreme hairiness is caused by an autosomal incompletely dominant gene termed N^d which is seen in the Drysdale, but another completely dominant autosomal allele also exists termed N^t which is not allelic to N^d, so this means that genetics are made more complex. Other mutations in this field may already exist or be produced in the future.

Unlike Australia, New Zealand and the USA, the British sheep industry is very stratified with hill sheep being selected for hardiness and bred pure. Draft hill ewes are then mated in the uplands to longwool breeds to improve size and fertility and these F1 crossbreds are then mated in the lowlands to Down breeds. Selection at each stage is different yet all three sections play a part in the final slaughter generation. About a third of the genes in the final slaughter carcasses stem from each of the Hill and Down breed groups. Selection is thus complicated. Additionally the hill farmer earns a higher percentage of income from wool than does the lowland farmer.

Down breeds are essentially terminal sires – what used to be called fat lamb sires but which might be better described as lean lamb or prime lamb sires. Within reason, fertility is less important in the terminal sire breeds and selection objectives are best directed at lamb growth rate or slaughter weight (LSW), lamb survival (LS) and carcass traits. An index might be directed towards:

$$LS + LSW + CL$$

where CL is some measure of carcass lean. Increasingly ultrasound is being used to assess lean. A programme is underway at Edinburgh University with a Suffolk flock in which selection is based upon lean growth rate.

The use of ram lambs has the advantage of reducing generation interval although they would be used prior to fleece weight being available. This is, however, of less importance and can be assessed later. The use of ewe lambs for breeding in their first autumn also helps to reduce generation interval. Young ewes of this kind will be less fertile than adults but there is good evidence to suggest that the higher the weight of the ewe lamb at mating, and the later its age, then the better the lambing performance. Some growth check might result from such early pregnancy but the effect on lifetime performance is likely to be minimal if well managed.

Some characteristics of sheep are relatively simply Mendelian traits, e.g. aspects of medullation and fleece colour. Other traits are of importance to different sections of the industry in different locations. Structural soundness is obviously of importance in breeding stock and some culling is going to be undertaken of sheep with functional problems of this kind (lameness, lost teeth, etc.). Some kind of disease resistance (e.g. to facial eczema in New Zealand) may be useful but it still needs to be fully understood in genetic terms. Hardiness is held to be highly important in Britain but difficult to assess, and it might be argued that looking at lamb productivity helps to assess this to some degree. In Australia/New Zealand woolly faces are undesirable and selected against, being held to have some importance in low fertility.

In evaluating growth traits of sheep, greater care is needed with correction factors than is needed with cattle. Birth type and rearing type are possibly the most important correction factors covering singles reared as singles, twins reared as singles and twins reared as twins. Between the first and last of these may be a 4.0 kg effect upon weaning weight.

Increased litter size is considered important and crossbreeding with Finnish Landrace was used in Britain as a means of increasing lamb number. This was successful but at the expense of growth rate such that Finn 'blood' is rarely used at more than 25% of crosses. It can be argued that unless management is at its peak, increased lambing potential will not be realised. Some association between testicle size and litter size has been suggested, but whether selection on testicle size will successfully advance litter size within breeds remains to be seen.

In this connection a fecundity gene (F) has been discovered in a strain of Merino called the Booroola. This was developed in New

South Wales, originally by the Sears brothers then further advanced by CSIRO. It is one of the more prolific strains of sheep as well as being an example of where a single gene can be of vital importance. It appears that FF sheep have an ovulation rate about three times that of normal (+ +) animals while heterozygotes (F+) are intermediate. At birth the differences are less obvious between FF and F+ animals but both excel over + + sheep. Work is being done to see how this Booroola F gene (as it is called) can be exploited for other breeds as well as the Merino.

Pig breeding

Pigs as litter producers have only recently begun to have considerable selection pressure placed upon numbers of pigs reared per litter and numbers of litters reared per year. Both need to be recorded along with weaning weights of each piglet, age at weaning and total weaning weight per litter. Although weakly inherited, litter traits tend to be quite repeatable (up to 30%) so that culling of poor performers will enhance herd means.

Physical defects, particularly leg weaknesses, are quite crucial in pigs, and sows and boars would be culled if defective in such features. It may be that there are more inherited anomalies in the pig than other farm species and such defects as cryptorchidism (very rare in cattle) are frequent in pigs along with anal atresia and other problems. It may also be that inbreeding in pigs is potentially more damaging than in some other species.

Because feed costs are a high proportion of total costs, there is great emphasis placed upon feed efficiency, and since pigs are usually housed and fed largely complete diets, feed intake is easier to measure than in other species. It is, however, essential to ensure that selection for increased feed efficiency is not achieved by the reduction of appetite as has been done in the past.

Pigs tend to be slaughtered at pork (65–70 kg), bacon (90 kg) and heavy hog (110 kg) weights, and at one time selection for bacon and pork pigs tended to be different. Now these aspects are usually run together. Although, as will be seen, there are basic differences in the carcass requirements and divisions may yet be resurrected.

Selection is usually based upon growth rate (possibly highly related to feed efficiency) together with ultrasonic measurements of fatness (selecting against backfat). Increasingly, testing is done on farm or more probably, within the units of commercial companies. Breeds are less distinct than was once the case with increasing differentiation into females lines (selected for reproductive traits) and male lines (selected on growth and carcass attributes). Most testing is to a fixed weight with backfat or, more rarely, eye muscle area assessed ultrasonically. In some countries there is preference for white pigs such that Large White/Landrace types are favoured, but some American breeds, developed primarily for pork, are coloured as are some of the European breeds.

Pigs are usually housed in groups and group size will have some effect upon performance. Individual housing, while it gives ready availability of intake data, is not only costly but contra-indicated as it might lead to genotype-environment interactions where progeny are to be group fed. There is also evidence that individually housed boars may develop libido problems in later life. Welfare aspects have also to be considered as pigs are social animals and derive some benefits from being in social groups.

Ideally, group housing with individual feeding is called for but this is only now becoming feasible as electronic feeders, long available for cattle, are becoming available for pigs. Without these feeders growth rate and lean production have to serve as indicators of feed efficiency. The increasing use in some countries of boars for meat production not only leads to greater efficiency of production but allows selection of males to be delayed until information becomes available. Thus boars and gilts that are above average can be selected from the bacon/pork pens.

Some carcass traits cannot be assessed ultrasonically either because instrumentation does not exist or it is too costly. This means that some progeny testing for certain carcass traits is called for. However pig breeders frequently use combined/sib selection by selecting boars and gilts from litters and carcass assessing some of these to give a guide to the potential of the boars retained. The British Meat and Livestock Commission (MLC) testing stations used to use two boars, a gilt and a castrate from each litter tested, killing the gilt and castrate to allow carcass data to be assessed for use in the boar index.

Single genes are rarely of major importance in many species but the discovery of the halothane gene has been important in the pig. Halothane is an anaesthetic which, when applied to some pigs at

Table 9.2 **Summary of halothane positive incidence (Webb, 1981)**

Breed*	Pigs tested	Halothane positive (%)
Duroc	248	0
British Large White	764	0
American Yorkshire	225	0
American Hampshire	232	2
Dutch Yorkshire	1394	3
Norwegian Landrace	576	5
Swiss Large White	1130	6
Danish Landrace	1990	7
German Landrace (GDR)	300	10
British Landrace	1538	11
Swiss Landrace	7480	13
Swedish Landrace	1668	15
Dutch Landrace	4073	22
French Pietrain	335	31
German Landrace (GFR)	1251	68
Belgian Landrace	1260	86
German Pietrain	266	87

*The review paper has more breeds given but only those with over 220 cases are reported here.

around 8–12 weeks of age, causes some of them to exhibit rigidity of the limbs. This is due to homozygosity for the recessive halothane gene. This gene appears to be associated with adverse effects such as stress susceptibility, pale soft exudative (PSE) meat and reduced litter size. There are, however, beneficial effects upon lean meat yield. Breeds appear to differ markedly in their incidence of the halothane gene and hence upon the incidence of its effects. Incidence figures are shown in Table 9.2 but clearly can be rapidly altered by selection.

Although there was some evidence that heterozygotes for the halothane gene were somewhat intermediary for all production features there is increasing evidence that there are still disadvantages. Since some of the better conformation breeds have higher halothane incidence this needs to be taken into account when producing or using synthetic 'meat' lines. There is a possibility that the chromosomal location of the halothane gene has been found in which event its elimination can be done more effectively. In general terms there are probably more disadvantages than advantages for this gene and it is probably best eliminated. There might be some future for it in a sire line that was heterozygous (Hh) for halothane (formed by crossing a non-halothane line HH with a homozygous halothane line hh) and using this on an HH female line. Slaughter progeny would be either HH or Hh and this could give a slight economic advantage but the future for the halothane gene does not look good.

In recent years the pig industry has gone the way of the poultry industry in being increasingly controlled (in breeding terms) by private companies, often internationally based. Although the consolidation of poultry into relatively few hands has been somewhat counter productive to the widespread understanding of poultry genetics the same may be less true of the pig industry. Pig breeding companies will, of economic necessity, be reluctant to publish everything that they are finding out about the pig but on the credit side there may be advantages that could outweigh disadvantages.

Companies are able to invest more than private breeders and they can guarantee greater health status as well as greater flexibility of objective. There is central control which can unify objectives even if those objectives are multiple, although the world interests of such companies may mean that the needs of a specific country are subservient to those of a wider market.

The fact remains that the old testing stations have gone and this may be no bad thing. At the time test stations were established their objective was to eliminate herd differences from sire evaluation. The use of BLUP techniques means that sires used in different herds in widely different locations can now be evaluated while still in those herds as long as reference sires are in use.

Figure 9.1 illustrates the type of set-up that has been established by some pig companies. Four units of 250 sows established in different locations for health/safety reasons are serviced by boars from an AI station run by the company. A number of boars would be used over a given period of time with each boar being used equally in each sow unit. Sire testing would be undertaken using BLUP techniques more effectively than was possible with the old test station procedure. Boars could be replaced several times per year thus making sure that inbreeding risks were minimised.

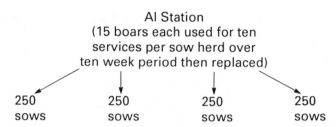

Fig. 9.1 **Schematic picture of a big breeding company scheme (gilts replace sows sequentially in each sow herd)**

In Fig. 9.1 there would be 1000 sows and with 15 boars, replaced five times per year, a total of 75 sire families. Sows would be replaced on a sequential culling basis, i.e. each sow is replaced as soon as a better gilt is found so that the better sows stay longer. Breeding values of sows would be updated at each litter. Investment is better directed at improved AI facilities than in test stations which are redundant because of BLUP.

Pig breeding companies can increasingly bring about genetic progress by the scale of operation and the objectives of that progress can more readily be modified. It is probable that sire lines will be selected for appetite and growth rate on *ad libitum* diets with a carcass lean content of around 58% to 62%. It may well be that selection against backfat thickness has gone far enough and as a result three problems have arisen.

Firstly there is the thin sow syndrome where females do not have enough fat cover during the lactational/breeding stage, and secondly selecting to reduce fat over the *longissimus dorsi* has reduced fat there only to see wedges of fat appear further down the rib cage and into the belly. Finally there is tissue separation after curing as the fat breaks away from the lean. Fat selection will thus have to be undertaken by measurements of a more extensive nature in various locations on the animal.

In female lines selection for litter size per sow and per sow per year will increasingly be important. Direct selection is not usually very successful because of the low heritability of the character with estimates of about 0.12 of a pig per litter per year. The French system of hyperprolific pigs in which outstanding litter producers (females) were selected from multiplier units and placed in nucleus units is more effective (0.30 pigs per sow per litter per year). The development of the family index at Edinburgh has been more useful still. In this system the potential of a gilt is assessed by looking at the performance (for litter size) of her mother and grandmothers as well as paternal and maternal half and full sisters. The advantage of such techniques means that the same procedure can also be done for young boars. This would have the advantage that selection for litter size could be undertaken in both sexes as opposed to merely the females. Estimates suggest that improvement increases approaching 0.50 pigs per sow per litter/year are feasible.

In the meat quality field it is possible that genetic improvement will be less effective than improvements in the abattoir and in processing. Eating quality is nevertheless receiving increasing attention. It is being argued that intramuscular fat is a principle component of eating quality. This has led to increasing interest in the Duroc breed of pig which has higher intramuscular fat levels. The breed certainly excels over most others when used for fresh pork where intramuscular fat levels in excess of 2% seem to be desirable. However for processing meat the breed does less well because high levels of intramuscular fat (over 3%) are undesirable. This may lead to different lines with some bred for the processing/bacon market and others aiming at the luxury fresh pork market.

Increasingly, too, the welfare lobby is making it economically desirable to produce pigs that are kept in less confined conditions. Selection of pigs in group housing is possible given feeding crates which will record the use of the crate by each individual pig as well as its intake. These exist but there are still problems with keeping pigs in group conditions. It could be that one is simply replacing one kind of stress seen in individual crates with a different kind of stress occasioned by 'pecking orders' and fighting. Outdoor pig production is another area of development for which different strains/breeds might be needed.

The use of Chinese breeds of pig is also something new. The breeds in which interest is expressed like the Meishan have ovulation rates in the region of 18 rather than 13, and litter sizes around 13 as opposed to 10. Chinese sows also have a larger number of teats. Unfortunately Chinese pigs are not only totally different in appearance from western pigs but are slow growing with very high carcass fat levels. Incorporation of the Chinese breeds into a maternal line is feasible if carcass/growth disadvantages can be minimised.

10 Breeding Policies in Developing Regions

Introduction

Although animal breeding technologies are widely accepted and fairly sophisticated throughout the developed world this is less true in what are generally called Third World countries. Such regions are frequently characterised by having tropical or sub-tropical environments although with considerable variations. Farming is either very fragmented at a subsistence level or very extensive, and even if animal agriculture is highly important the sociological and political situation may be quite different from Europe or North America.

The major animal problems in such regions are usually concerned with management and nutrition. Breeding livestock is regarded as a complex business involving various disciplines and there is no general acceptance that genetic improvement would be useful, particularly in subsistence systems. Developing nations in Latin America, Africa and Asia have about 80% of the world's human population, some 60% of the world's cattle, a third of the world's sheep and very high proportions of the world's buffalo. Despite the high proportions of cattle these regions provide a small proportion of the milk and meat produced in the world. Increasing the efficiency of animal agriculture, especially that of cattle and other ruminants, which are in less direct competition with man than are pigs and poultry, would clearly be beneficial to these regions. The difficulty arises in transferring sophisticated breeding schemes devised for the developed world into the developing regions. Highly commercial and highly specialised systems may not be biologically, economically or socially desirable or appropriate for the developing world.

Dairy cattle

In developing nations which are frequently tropical, ranging over wide ecological zones, cattle breeds have been developed which are usually of *Bos indicus* as opposed to *Bos taurus* origin. Such cattle are characterised not only by different morphological features such as large drooping ears, pronounced dewlaps and humps but they are frequently better suited to the areas concerned. They are more climate adapted, having different sweat glands, and are frequently more tolerant to enzootic diseases. However much of their 'adaptability' may stem from their lower productive performance.

There is some evidence that tropical breeds show much greater variation in milk yield than European breeds and that maternal instincts towards the calf are more pronounced with consequent difficulties about milk let down unless the calf is present.

Although selection has been undertaken among tropical dairy cattle, such studies have tended to be either based upon small numbers or undertaken over too short a period of time. Genetic variation certainly exists among such cattle for many economically important traits, but selection within native breeds which are late maturing and low producers of milk is likely to be unrewarding.

On the other hand, replacement with European stock is fraught with dangers. Imported European breeds can be successfully farmed in the tropics when there is substantial control of disease, and feeding and management are high, with climatic stress reduced to a minimum. Such conditions are likely to prevail only in special circumstances and, in the main, European imports are likely to exhibit susceptibility to disease, reduced growth, poor fertility and high mortality (see review by Pearson de Vaccaro, 1990). Replacement techniques are likely to succeed only in special circumstances when linked to improved management skills. These will, of course, apply in some places.

Replacement can be undertaken by the use of semen and AI which means not only that the initial introduction is relatively cheap but that the first produce are crosses between native cows and semen of European-type bulls. The difficulties with AI are that in developing nations the techniques are not nearly as advanced as they are in Europe with inadequate levels of training for AI personnel. The importation of semen requires hard currency which may be a stumbling block. In addition, roads and communications may be

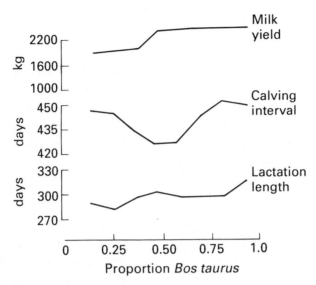

Fig. 10.1   **Performance of different *Bos taurus/Bos indicus* combinations
(after Cunningham & Syrstad, 1987)**

poor and there may be defective nutrition and disease among cattle.

In a review of the subject Cunningham & Syrstad (1987) showed
that introduction of up to 50% *Bos taurus* genes into *Bos indicus*
cattle tended to improve performance in most traits but beyond this
point calving interval increased (due to reduced fertility) while most
other traits showed few clear trends (see Fig. 10.1).

Any crossbreeding programme needs to be correctly evaluated
which means that not just one feature has to be considered. The
crossing of Sahiwal with Friesian might give an increased milk yield
even if taken beyond the 50% Friesian level but continued upgrading
might bring with it longer calving intervals, more reproductive
disorders, increased disease susceptibility as well as disadvantages in
fringe areas like meat production or draught capability. The success
of the programme has to be assessed in overall terms rather than on
the basis of one trait like milk yield. Pearson de Vaccaro's (1990)
study of survival rates showed imported breeds to average 2.6
calvings versus 3.1 for local breeds. However better results were
seen from crossbreeds containing 50% to 63% European 'blood'.

The specialist dairy herd in Europe or North America is exactly
that – *specialist* – but in the tropics the dairy cow may be used not
only as a provider of milk but also for beef through her calf, draught,

and as a source of fertiliser through her dung. In some circumstances productivity is not easy to assess especially if, as in parts of Africa, the herd is a nomadic one.

In some areas of Latin America it is policy to milk the cow once or twice daily while allowing the calf access to the cow after each milking. The cow thus provides milk as well as rearing her calf through to around six months of age. Therefore, definition of breeding objectives is essential and would be quite distinct from that in European/North America.

It does appear that heterosis between native and exotic cattle is greater in poor, rather than good, environments but that heterosis is likely to be higher between *Bos taurus* and *Bos indicus* types than that between two *Bos taurus* breeds. Most experts would argue that some influx of exotic 'blood' is needed but without a hard and fast rule being possible as to the degree of inclusion. AI might be the best way of incorporating exotic breeds but AI is not necessarily easy for the reasons already given.

Small herd size makes organised schemes virtually impossible. There is a need for government involvement aiding recording, and AI, or even assisting in the setting up of a nucleus type unit from which might come the semen or the breeding bulls.

In good environments *Bos taurus* breeds might be the best choice either by importation or grading up. In less good situations the choice may be restricted to the degree of exotic 'blood' incorporated either by a rotational crossing system or by new breed production as was done to form the Siboney de Cuba (see Chapter 8).

With the dual purpose system of milk plus suckled calf tne breed type will probably be a crossbred rather than a straight exotic. Such systems are often more successful in the calf rearing field than artificial rearing from a few days of birth. Objectives may be readily defined, e.g. 1800 kg of milk plus a 180 kg weaned calf plus a fertility level of 85% but selection for such a composite is not easy.

Beef cattle

Beef cattle in tropical areas are an easier proposition than are dairy cattle. Again beef cattle in these regions tend to be of zebu type (*Bos indicus*) although in parts of Latin America breeds exist that are of *Bos taurus* origin but which have been there since the time of the *conquistadores*. Under the generic title of Criollo these include such breeds as the Blanco Orejinegro (BON), the Romosinuano and the Costeño con Cuernos. Although European in origin these breeds are virtually indigenous by long association with the region. Criollo cattle may be more fertile than Zebu types which are unlikely to present oestrus much before 22–24 months of age while temperate breeds will do so around 15–16 months even in tropical conditions. In Mexico the term criollo is often used to indicate 'mixed' blood cattle rather than the breed types previously mentioned.

Beef cattle production levels tend to be poor with calf crops around the 40–60% level compared with 85–95% in Europe. Seasonality of herbage does not aid the situation. Calving intervals are closer to 460 days than the 380 days of Europe albeit with a wide range around this figure. Age at first calving is likely to be closer to three years in imported *Bos taurus* cattle and closer to four years in native types. Individual herds can do very much better than average in all these traits but such units are few and far between and although exotic breeds can drastically reduce age at first calving, as well as calving interval, the reduction in the latter is largely true of the first interval and less so of subsequent ones.

Dystocia is rarely a problem in native breeds but many of them produce light calves (around 28 kg at birth) and it is well established that mortality is higher is small calves than in more standard sized ones (about 36 kg). Crossing with European breeds will increase birth weight but also increase dystocia risks to some degree. Tropical breeds tend to be more difficult to get in calf while lactating than when dry which does not help reproduction. Mortality is high in most tropical cattle (minimum values of 10% to weaning at seven or eight months) but temperate breeds are not necessarily always better than native ones.

In many situations there is a need to look at the potential of native breeds before seeking to import exotic ones. In Southern and Eastern Africa the Africander was widely used in preference to local breeds but under scientific scrutiny indigenous breeds like the Tuli and Mashona have performed better (see Hetzel, 1988). The Brahman has been advocated to improve African cattle but justification from such trials as exist is hard to find.

Beef cattle breeding in tropical situations can be done by selection within local breeds using similar traits to those in Europe but because reproductive rates can be so very different the need to look at kg of calf weaned per cow mated is crucial. There is evidence that temperate breeds of *Bos taurus* origin as well as Brahmans can be effective as crossing sires and there is thus a need to study crosses between the better types of indigenous cattle like the Tuli and Mashona with exotic breeds. Such evaluations need to be undertaken under the local conditions where such cattle will be farmed.

In Latin America Plasse (1983) showed advantages from crossbreeding with exotic breeds on native Zebu cattle but less advantage from crossing Criollo breeds with Zebu types beyond the first F1 generation. There may be evidence to suggest better results from European breeds than those of British origin but more data are needed. However rotational crossing does seem to be advantageous (see Chapter 8).

Pigs

The pig is perhaps better adapted to tropical conditions than most farm species and there are numerous breeds of pig which have been developed in tropical regions. Most of the criteria thought desirable in temperate regions would be equally acceptable in tropical environments. One would thus be seeking large litter size, frequent farrowing, survival ability and rapid lean growth. Some pig breeds such as those in China are extraordinarily prolific but in keeping with many of the pig breeds of Central and South America they tend to be lard (fat) breeds with undesirable carcass traits.

It is probable that most breeds will express themselves best in the environment in which they were selected, especially since genotype–environment interactions are more likely in pigs than, say, cattle.

Imported mature stock are likely to suffer in the tropics and to be less fertile at least for some time. Boars may show a reduced libido and females produce either smaller litters or show increased mortality especially if, as seems likely, milk yield is depressed.

On the other hand carcass improvements will be better achieved by the use of imported temperate breeds and a policy of grading-up seems called for using imported stock brought in at young ages. As such a programme continues, acclimatisation will be improving and performance will be increasingly better.

Disease will be a major drawback in tropical regions and the ability to produce disease-free strains as in temperate regions will be less because of the cost factor. Simple procedures could help to reduce disease risks to some degree and aid in selection by producing more stock available for that selection.

In general breeding aims and their achievement will be similar to those in developed regions.

Sheep

Sheep are not ideally suited to tropical regions partly because wool is to some degree disadvantageous in such regions, but also because of photoperiodicity. Sheep in the tropics, especially imported stock, have some difficulty with reproduction when day length is fairly constant. Adapted sheep breeds do cope well with the lack of major day length changes and some tropical breeds (e.g. Barbados Black Belly) are highly prolific. Many tropical sheep breeds are hair sheep rather than wool producers, and most tropical sheep tend to be small.

Selection criteria in the tropics are likely to be similar to those in temperate regions but if large sheep are to result some upgrading with temperate breeds seems necessary.

11 Breeding Schemes and Future Developments

Introduction

Few animal breeders operate on a very large scale. The average size of dairy herds varies throughout the world but in most locations would be less than 50 cows. Pedigree beef herds in Europe would be closer to 10 or 20 cows though they might be very much larger in ranching situations in the Americas or in Australia. Pig herds may, in many cases, exceed 100 sows while sheep flocks can be numbered in thousands, especially in hill situations in Britain, and even larger in Australasia where numbers may run into many thousands. Nevertheless, many important breeds are found in quite small units. In Britain, for example, most longwool flocks (important in the stratification of sheep breeds) would be no more than a dozen ewes. Many breeders of working sheepdogs might have a larger breeding unit!

Because genetics is essentially a numbers game the relatively small size of breeding units places great restriction on the progress any single breeder can expect to make. In most species the genetic progress expected stems, in large part, from the fact that government or semi-government bodies have become involved as well as commercial companies. Poultry breeding is almost exclusively in the hands of large commercial companies and most producers simply buy day-old chicks from the company of their choice, replacing them with others once they have been reared to slaughter or laid eggs for the required cycle. Increasingly pig breeders are following this example by purchasing gilts/boars from private companies which control the breeding programmes. Cattle and sheep breeders have still remained outside of this sphere of influence although organisations controlling AI such as Dairy Boards are increasingly controlling the sire end of the dairy and beef market and may do the same for sheep once AI becomes more efficient and acceptable in that species.

Group breeding schemes

Increasingly the private breeder is being forced to consider cooperative ventures of one kind or another. While pedigree breeders may retain the element of commercial and breeding competition with their colleagues they are increasingly aware that cooperative schemes, either through their breed societies or in collaboration with fellow breeders, are becoming more necessary. One result has been the development of group breeding schemes which started in Australia/New Zealand.

Both sheep and beef cattle have benefited from such schemes, often called nucleus schemes. They date back to the New Zealand Romney scheme of 1967 and have expanded since then into some quite large units.

Essentially group schemes involve a series of breeders who agree to cooperate using their own flocks/herds which can be pedigree, commercial, or a mixture of the two. Usually a nucleus unit is established on either a separate unit or on one of the collaborating farms. This nucleus unit is usually formed by selecting the 'best' females from each of the cooperating units. What exactly 'best' means will depend upon the criteria used by the particular scheme. If only pedigree stock are used there will be a two-tier system but with the addition of commercial animals this becomes a three-tier system.

Groups can vary enormously in both size and objectives but the essential format involves quite intense recording and measurement in the nucleus unit. The very best males produced are used in this unit and replacement females also stem, in part, from this unit. In addition, some females are sent to this unit each year from the cooperating farms. The cooperating farms receive males from the nucleus unit. Because there is a two-way flow of stock (and genes) both into and out of the nucleus unit it is termed an open nucleus. An example of a group breeding scheme is shown in Fig. 11.1 using hypothetical figures.

In general, nucleus units contain about 10% of the total size of cooperating units and take about half their replacement females from these cooperators each year. It is generally considered that

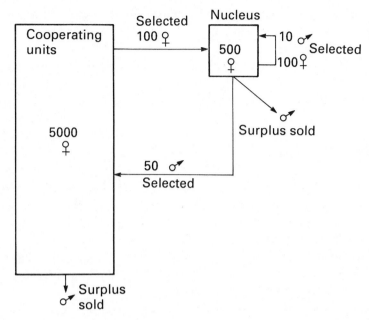

Fig. 11.1   **Group breeding scheme**

such units can produce 10% to 20% greater progress than could individual breeders operating alone.

Such schemes have the advantage of continuity (they can exist even if one or two cooperators withdraw), group identity and, by virtue of larger financial size, they can provide improved management, technical assistance and customer services while at the same time providing better data collection, shorter generations, index selection and the like.

On the debit side the nucleus unit is potentially subject to disease problems because of constant input of new stock, there are some difficulties about assessing genetic merit in the different cooperating units, and there is the danger of trying to screen the incoming females for too many features, which adds to potential cost. Closed nucleus units would avoid the health risk to a degree but should make less theoretical progress as a consequence.

Despite their success in New Zealand and Australia such schemes have been rare in Britain. A Welsh Mountain sheep unit has operated for some years now and a kind of nucleus unit exists in Edinburgh for Suffolks. In France the hyperprolific pig scheme is a kind of nucleus unit into which the most prolific sows from multiplier units are sent. Here, however, one has to leave time for the genes from the influx to fill the nucleus and then filter back to the multipliers before a further selection can be undertaken. Progress is faster than simple selection for litter size but it can only be undertaken about every three years or so.

Artifical Insemination (AI)

By no means a new technique since it traces back to the 19th century and has been in commercial use from the 1940s, AI has been a major tool in the advancement of dairy cattle. Not only has it helped in the more effective testing of dairy sires and been instrumental in the development of contemporary comparison techniques but it has also aided the changing of breeds by migration/grading up, not only within a country but between countries.

Nevertheless it has not been as widely used in beef herds for obvious practical reasons connected with the more extensive nature of beef production and/or the small size of some herds. In Britain beef AI has had a major impact in the spread of new breeds but within some pedigree breeds registrations rules have sometimes been a positive dampener on the spread of AI and have made the use of superior sires to produce future sons not always easy. To be widely applied with beef cattle, effective and cheap synchronisation is needed.

This has been feasible in sheep but here the disadvantage has been the difficulty of freezing semen coupled with some inability to actually identify the superior sheep sires that would justify sheep AI. Pig AI is certainly functioning, albeit better in some locations than others, but here again there is room for improvement before AI can emulate in other species its success in dairy herds. Fortunately at the level of pig companies AI is very important and essential to the BLUP techniques being used.

The sexing of semen will be a step forward, principally in cattle, but, despite the years of searching for new technologies to achieve this, the current success rate is rather illusory. Not only would sexed semen allow breeders to produce the desired sexes from the chosen females but also it would allow for the easy use of single sex beef systems (once bred heifers). In pig units sexing would allow breeding companies to produce only females in the sow lines and only males in the sire lines, aside from replacements. Boars from sow lines are of no value other than for fattening and even then they are less useful than the end products of sire/dam line crosses.

Embryo transfer

The role of MOET schemes in dairy production has already been mentioned (see Chapter 9) and as the techniques becomes easier and cheaper and the success rate improves increased reproductive rates for the 'best' cows will add to progress from the female side. Sexing of embryos will enable breeders to determine what sex of calf they need, which could lead to an increase in females available for selection in a dairy unit, or an increase of male calves in a beef unit producing slaughter generation cattle. Embryo splitting means that pregnancy rates can be doubled (e.g. from 60% to 120%). At present two or three calves per embryo may be the limit but the future may see this increased. This can lead to increased twinning without the problem of freemartins seen in cattle with twins of different sexes. Embryo transfer does not, however, offer much to pig breeders.

Embryo research in the human field has been instrumental in helping couples with fertility problems as well as aiding in our understanding of inherited defects. At the time of writing political limitations on such research are being considered mainly because of lobbying by groups concerned with opposition to abortions. It is not outside the realms of possibility that opposition to embryo research in animals might follow and much potential progress might be terminated.

Gene transfer

The ability to transfer genes across species may appear to offer many opportunities. In reality the impact upon large animal breeding may be more imagined than real. Transgenic animals were only produced as recently as 1980 so that the technology is still in its infancy. Most work has been undertaken in species like the mouse which is of minimal interest to the livestock breeder. If gene transfer is to be of value we must first:

(1) Identify major genes that are useful.
(2) Locate these genes in the donor.
(3) Transfer the gene successfully.

The problem is that as far as (1) is concerned few major genes are known that are important. The ability to identify genes involved in hormone production has allowed the production of sheep which produce such products as insulin for human use. This is undoubtedly valuable and new generations of proteins for pharmaceutical use in humans or other animals are likely to ensue but this hardly constitutes an advance in farming.

Genes that might be useful would be the muscular hypertrophy gene in Belgian Blue cattle or the fertility gene of Booroola sheep but as yet these have not been located, still less transferred. Advantages of the halothane gene in the pig are more questionable.

Transfer of genes will only be successful if there are no side effects. It may be academically interesting to create mice the size of rats but in agricultural livestock this success has not been repeated. When genes have been transferred as with growth genes to pigs there have usually been side effects connected with arthritis or sterility. Thus even transferring the gene once located is not the end of the story. However progress in this area may well result.

It may well be that macro techniques of the traditional animal breeder will be needed for a long while yet quite apart from the fact that genetic engineering may fall foul of political decisions and have many curbs placed upon its use.

At a time when agriculture, in Britain especially, is increasingly coming under pressure from the 'environmental lobby' (not always wisely or fairly) there is real possibility that gene transfer will antagonise that lobby still further. It may be retrograde and unfair but 'lobbies' rarely seek to examine both sides of an argument. The geep (goat/sheep hybrid) may sound or look interesting but it is likely to arouse the ire of the layman whose political clout is far greater than that of the scientist.

We might end up deciding that the idea of gene transfer is as Dr John King said (1988) 'a false dawn which opens up new opportunities yet to be realised'. It is with regret that I share his pessimism.

Preserving genetic variation

Many breeds of farm livestock have declined in importance and some have become extinct. Usually this has occurred because populations were always small and struggled to survive in the face of inbreeding. Some have been popular breeds which have lost their popularity as criteria have changed and other breeds have been shown to be more economically viable for the needs of a particular era.

Some organisations have sought to preserve these rare breeds by farming them while others have sought to store semen or embryos; other breeds are preserved in zoos. The Food and Agriculture Organisation of the United Nations (FAO) has begun to catalogue such breeds.

It is difficult to preserve small groups because of inbreeding risks and disease threats which could wipe them out. Moreoever, if these rare breeds are less commercially viable than the breeds which have replaced them then those farming rare breeds do so at an economic disadvantage. Some argue that these out-of-date breeds serve no useful purpose while others argue that as demands change some of these breeds will be found to have valuable features that will one day be needed.

It is certainly true that in Britain breeds like the Jacob sheep have had a new lease of life in the general climate towards what is being called sustainable agriculture where there is something of regression towards non-intensive farming. This will help preserve some rare breeds but is unlikely to bring such breeds back to prominence. Unless a breed can be shown to have advantages over others and that these advantages have economic merit there is little future in rare breeds. Most declined for economic reasons and a rebirth is only going to occur for new economic reasons. Rare breeds will simply get rarer and eventually die unless there is government help to preserve them and any justification for this is open to argument.

Appendix 1
Coefficient of Inbreeding

The coefficient of inbreeding was devised by Sewell Wright in the 1920s and is an attempt to measure the degree of homozygosity in an individual. The formula is:

$$F_x = \Sigma 1/2^{n_1+n_2+1} \, (1 + F_A)$$

where F_x is the inbreeding coefficient, n_1 is the number of intervening generations from the sire to the common ancestor and n_2 is the number of intervening generations from the common ancestor to the dam. F_A is the inbreeding coefficient of the common ancestor.

Most textbooks use this formula and recommend arrow diagrams to construct pedigrees. This is acceptable as long as pedigrees are not complex but with complex pedigrees arrow diagram construction can lead to many errors. It is thus better to use the technique of Willis (1968) which alters the formula to:

$$F_x = \Sigma 1/2^{n_1+n_2-1} \, (1 + F_A)$$

where n_1 and n_2 represent the actual generations counting the parents as generation 1.

Using the pedigree of Roan Gauntlet (Chapter 6) we can describe this bull as inbred:

> Champion of England 3.3/2
> Lord Raglan 4.4/4

This means that Champion of England appears twice in the third generation of the sire's side and once in the second on the dam's side. Similarly Lord Raglan appears twice in the fourth generation on the sire's side and once in the fourth on the dam's.

The inbreeding to Champion of England is

$$1/2^{3+2-1} = 1/2^4$$

$$\text{plus } 1/2^{3+2-1} = 1/2^4$$

which is $0.0625 + 0.0625 = 0.125$ or 12.5%. Since there is no knowledge of the actual inbreeding of Champion of England the part $(1 + F_A)$ is ignored.

The inbreeding to Lord Raglan is

$$1/2^{4+4-1} = 1/2^7$$

$$\text{plus } 1/2^{4+4-1} = 1/2^7$$

which is $0.0078 + 0.0078 = 0.0156$ or 1.56% and again we ignore the $(1 + F_A)$ portion.

The total inbreeding is thus $0.125 + 0.0156$ or 0.1406 (14.06%).

If Champion of England had been inbred let us say 11.5% then the inbreeding to that bull of 0.125 would need to be multiplied by $1 + 0.115$ (1.115) and thus would have been 0.1394 or 13.94%. Note that in multiplication the proportion value should be used, not the percentage value. Multiplying by $1 + 11.5$ would result in a value over unity and coefficients cannot exceed 1.00 or 100% and in livestock would rarely approach this level. The common ancestor inbreeding is not usually important unless the common ancestor is inbred around 10% or more.

The advantage of the above technique is that there is no requirement to rewrite the pedigree which is used exactly as presented. The risk of confusion drawing complex arrow diagrams is also avoided. Care has to be taken when inbreeding is to a specific ancestor with secondary inbreeding to animals behind this ancestor. Ways of dealing with this appear in the original paper.

Appendix 2
Coefficient of Relationship

The simplest way to construct a coefficient of relationship between two animals (A and B) is to calculate the coefficient of inbreeding (as explained in Appendix 1) that would result from mating A and B and then doubling this figure. It does not matter if A and B are of the same sex since you are only calculating the theoretical inbreeding not actually making a mating. If A and B mated together gives an inbreeding coefficient of 0.079 then the coefficient of relationship would be 0.158 or 15.8%.

Glossary

Ad libitum: unrestricted.

Additive: that part of the variance which can be transmitted to the next generation.

Allele: any of the alternative forms of a gene.

Artificial Insemination (AI): the collection of male sperm and its placing in the female reproduction tract.

Artificial selection: selection made by man.

Assortative mating: mating animals which look or perform the same.

Autosomes: those chromosomes which are not sex chromosomes.

Backcross: cross between an F1 and either parent.

Biometrics: statistics as related to biological data.

Birth type: whether born a single or twin, etc.

Bloodline: usually refers to animals with some degree of relationship.

Breeding value: the merit of an animal as a breeding prospect.

Castration: removal of the tests. (Such animals termed *Castrates*)

Cell: basic unit of living organisms.

Chromatids: those parts of the chromosome after it has split longitudinally.

Chromosome: thread-like structures found in the cell nucleus on which genes are carried.

Coefficient of variation: statistical term. Standard deviation expressed as percentage of the mean.

Common ancestor: an individual appearing on both sides (sire and dam) of a pedigree.

Compensatory growth: growth occurring during a period of unrestriction after a period of restriction.

Contemporary(ies): animal born and reared at the same time (more or less) as those being studied.

Control(s): an unselected population against which selected ones are compared.

Correction factors: amounts added to a trait to make it comparable to other measurements. For example, to correct weight in a female to what it would have been as a male. Sometimes multiplication factors are used rather than additive ones.

Correlated response: effect on one character of selection undertaken for another.

Correlation: statistical term (r) describing extent of an association between two traits.

Crossbred: an animal produced from parents of different breeds.

Crossing-over: the exchange of parts of one chromosome with its homologous partner.

Cryptorchid: failure of a testicle to descend (unilateral) or both testicles to descend (bilateral) into the scrotum.

Culling: removing poor or unsuitable animals from the population.

Deleterious gene: gene (usually simple) causing the animal to be defective in some way.

Deoxyribonucleic acid (DNA): the chemical that is the basis of the gene.

Deviation: statistical term indicating the difference between an observation and the mean of the population.

Diploid: the normal number of chromosomes for a species.

Distribution: statistical term describing the spread or range in observations.

Docking: removal of part of an animal's tail.

Dominant: refers to a gene which operates when present only in single dose.

Dressing percentage: carcass weight as a percentage of the live animal weight.

Dropsy: an inherited defect where tissues fill with fluid.

Dystocia: difficulty in giving birth.

Embryo: the early stage of an organism in the uterus.

Embryo transfer (ET): technique involving removal of an embryo from one female (the donor) and placing it into another (recipient).

Enzymes: substances which trigger off reactions but are unchanged by those reactions.

Epistasis: interaction between genes (alleles) at different loci.

Fecundity: number of progeny born and reared.

Fertility: ability to conceive (female) or produce viable sperm (male).

Foetus: unborn young still in the uterus.

Follicle: structure in the skin from which hair fibres grow.

Fitness traits: those concerned with reproduction and viability.

Frequency: statistical term indicating number of times an observation occurs.

F1: first fillial generation (first cross).

F2, F3, F4: etc.: subsequent generations: e.g. F2 is F1 × F1.

Gamete: reproductive cells (ova, sperm) which unite to form the zygote.

Gene: the basic unit of inheritance.

Generation interval: average parental age when progeny are born.

Genetic drift: changes in gene frequency as a result of chance or random effects.

Genetic engineering: modification of an animal's genetic constitution by manipulation of genes.

Genetic gain: heritability/selection differential.

Genetic isolate: strain or line that is distinct from others.

Genotype: the genetic make-up of an animal.

Genotype–environment interaction: change of ranking order of different genotypes in different environmental conditions.

Germ cell: gamete.

Germ plasm: genetic material in an animal.

Haemophilia: Blood disease(s) affecting clotting time.

Half-bred: first cross (F1) between two different breeds.

Haploid: half the number of chromosomes typical for the species. This number is seen in eggs or sperm.

Heritability: h^2. The additive variance as a proportion of the total variance.

Heterosis: superiority of a cross over the mid-parent.

Heterozygote: an animal with unlike alleles at a locus (e.g. Bb).

Hogg (hogget): sheep aged 6 to 12 months of age (approximately).

Homologous: of common origin. Chromosomes of a pair.

Homozygote: an animal with like alleles at a locus (e.g. BB, bb).

Hormone: secretion from special glands which permits or encourages certain functions.

Inbreeding coefficient: the rate at which homozygosity is increased.

Inbreeding depression: decline in performance due to inbreeding.

Joining: putting males with females with a view to their mating.

Karotyping: examination of chromosomes.

Kemp: coarse, heavily medullated fibres seen in a fleece.

Killing out percentage: see *dressing percentage*.

Let-down: act of releasing milk from the udder, controlled by oxytocin.

Lethal gene: a gene which when expressed causes death.

Linkage: genes associated because they appear on the same chromosome.

Locus (plural *loci*): location on a chromosome for a specific gene.

Maintenance: feed required for an animal's basic functions when not performing a productive function like growth/lactation/reproduction.

Mean: total value of the observations divided by the number of observations.

Medulla: cavity or hollow inside a fibre of sheep hair.

Meiosis: reduction division from diploid to haploid state.

Migration: introduction of animals from one population (or breed) into another.

Mitosis: cell division to form two new but identical cells.

MOET: multiple ovulation and embryo transfer.

Monozygous: originating from the same egg.

Mutation: change in genetic material induced naturally or by certain chemicals or by radiation.

Natural selection: selection brought about by nature.

Nicking: used when a specific cross appears to be successful.

Normal distribution: bell-shaped curve which describes variation seen in polygenic traits.

Nucleus: centre of a cell in which the chromosomes are found.

Objective trait: one that can be defined and measured with some precision.

Oestrus: heat period in a female during which mating can occur.

Ovary: female reproductive organ.

Overdominance: when the heterozygote outperforms both homozygotes.

Parity: order of birth, e.g. first calving, second calving.

Pedigree: record of the ancestors of an individual.

Performance test: assessment of the animal's own performance.

Perinatal: around birth.

Phenotype: outward expression of the animal's genetic make-up.

Placenta (afterbirth): structure in uterus which surrounds the foetus and through which it is fed. Expelled after birth.

Plateau: term used to describe state where selection ceases to produce further progress.

Pleiotropy: gene having an effect upon more than one trait.

Polygenic: controlled by many genes.

Population: group of individual animals.

Prepotency: ability of an animal to reproduce its own features.

Progeny test: evaluation of an animal on performance of its offspring.

Puberty: sexual maturity.

Qualitative traits: traits that are distinct or discrete and can be counted, rather than measured. Often simply inherited.

Quantitative traits: traits that are measured and which show continuous variation. Usually polygenically inherited.

Random: by chance.

Recessive: allele that has to be present in duplicate to be obvious. One that is masked by a dominant allele.

Reciprocal cross: cross made in both directions, e.g. Charolais male to Hereford female and vice versa.

Reduction division: see *Meiosis.*

Regression: statistical term (b) measuring by how much one trait changes for each unit change in another trait.

Relative Economic Value (REV): estimate of value (financial) of one trait relative to others.

Repeatability (R): extent to which a trait is repeated next time around.

S/P ratio: ratio of secondary to primary follicles in wool (measure of quality).

Segregation: separation of alleles of a pair to form germ cells.

Selection differential: difference between mean of parents and mean of population from which they were drawn.

Selection index: selection for multiple objectives by pooling information into a single score or index.

Selection intensity: severity with which selection is practised, e.g. proportion used for breeding. Mathematically the selection differential/phenotypic standard deviation.

Semen: male sperm and lubricating fluids produced by the testes.

Sex chromosomes: that pair of chromosomes determining sex (XX and XY in animals).

Sex controlled (influenced): condition seen more frequently in one sex than the other.

Sex-limited: condition expressed only in one sex.

Sex-linked: condition carried on the sex chromosomes.

Sibling (sib): brothers and sisters. Full sibs have same parents, half sibs have one parent in common. May not be from same litter.

Skewed: distribution that is distorted with a long tail to one side.

Standard deviation: statistical term describing range of variation around a mean. Square root of the variance.

Stud: term used to describe a herd/flock of pedigree animals usually of some fame (Australiasian term). In Britain used to describe a male offered for breeding (mainly canine term).

Subjective trait: one defined in a non-precise way usually assessed by a score or on an hedonic scale.

Superovulation: stimulation of the ovary to produce more eggs than normal.

Teaser: vasectomised male used to detect females in oestrus.

Telegony: discredited belief that crossbreeding a female will affect all her future offspring from future matings.

Test-mating: mating a suspect carrier of a deleterious gene to an animal known to have that gene in order to check genetic status of the testee.

Threshold trait: one which is seen in limited forms (usually two or three) but which is polygenically controlled.

Top-cross: using a sire of the same breed but new bloodlines.

Trait: characteristic or feature of an animal.

Truncation point: performance level at which selection is made and some animals culled.

Type traits: those associated with physical features relating to a standard of excellence usually drawn up by a breed organisation.

Ultrasonics: equipment that identifies fat layer, muscle area or foetus, using high frequency sound waves.

Uterus: female organ in which foetus develops.

Variance: statistical term to measure variation in a population. Square of the standard deviation.

Yearling: animal of about 12 months of age but under 24 months.

Zygote: product of union of two gametes.

References

Becker, W. (1984). *A manual of quantitative genetics*, 4th edn Academic Enteprises, Pullman, Washington.

Cunningham, E.P. & Syrstad, O. (1987) Crossbreeding *Bos indicus* and *Bos taurus* for milk production in the tropics. *FAO Animal Production & Health Paper* No. 68. 90 pp.

Dalton, D.C. (1985) *An Introduction to Practical Animal Breeding* 2nd edn. BSP Professional Books, Oxford.

Falconer, D.S. (1989) *Introduction to Quantitative Genetics* 3rd edn. Longman, Harlow.

Gibson, J.P. (1987) The options and prospects for genetically altering milk composition in dairy cattle. *Anim. Breed. Abstr.* **55**, 231–43.

Henderson, C.R. (1949) Estimation of genetic changes in herd environment. *J. Dairy Sci.* **32**, 706 Abstr.

Henderson, C.R. (1973) Sire evaluation and genetic trends. In *Animal Breeding & Genetics*. Am. Soc. Anim. Sci./Am. Dairy Sci. Assoc., Champaign, Illinois.

Hetzel, D.J.S. (1988) Comparative productivity of the Brahman and some indigenous Sanga and *Bos indicus* breeds of East and Southern Africa. *Anim. Breed. Abstr.* **56**, 243–55.

King, J.W.B. (1988) The future role of the new technologies–what opportunities do they offer? *Conference on Harnessing the New Technologies for profitable beef breeding & production.* Stoneleigh.

Lamberson, W.R. & Thomas, D.L. (1984) Effects of inbreeding in sheep: a review. *Anim. Breed. Abstr.* **52**, 287–97.

Lauprecht, E. (1961) Production of a population with equal frequencies of genes from three parental sources. *J. Anim. Sci.* **20**, 426–32, also errata p. 902.

MMB. (1987) *Report of the breeding and production organisation.* No. 37. Milk Marketing Board, Thames Ditton.

Newton Turner, H. & Young S.S.Y. (1969) *Quantitative genetics in sheep breeding.* Macmillan, Melbourne.

Nicholas, F.W. (1979) The genetic implications of multiple ovulation and embryo transfer in small dairy herds. *Proc. Conf. EAAP,* Harrogate.

Nicholas, F.W. (1987) *Veterinary Genetics.* Clarendon Press, Oxford.

Nicholas, F.W. & Smith, C. (1983) Increased rates of genetic change in dairy cattle by embryo transfer and splitting. *Anim. Prod.* **36,** 341–53.

Pearson, K. (1931) *Tables for statisticians and biometricians Part 11.* Biometric Laboratory, University College, London.

Pearson de Vaccaro, L. (1990) Survival of European dairy breeds and their crosses with Zebus in the Tropics. *Anim. Breed. Abstr.* **58,** 475–94

Plasse, D. (1983) Crossbreeding results from beef cattle in the Latin American tropics. *Anim. Breed. Abstr.* **51,** 779–97.

Ponzoni, R.W. & Newman, S. (1989) Developing breeding objectives for Australian beef cattle production. *Anim. Prod.* **49,** 35–47.

Preston, T.R. & Willis, M.B. (1976) *Intensive beef production* 2nd edn. Pergamon Press, Oxford.

Sheridan, A.K. (1988) Agreement between estimated and realised genetic parameters. *Anim. Breed. Abstr.* **56,** 877–89.

Simm, G., Smith, C. & Prescott, J.H.D. (1986) Selection indices to improve the efficiency of lean meat production in cattle. *Anim. Prod.* **42,** 183–93.

Webb, A.J. (1981) The halothane story in pigs–lessons for poultry. *33rd Poultry Round Table,* Edinburgh.

Willis, M.B. (1968) *Revista cubana Ciencia Agricola* (English edition) **2,** 171–4.

Willis, M.B. (1989) *Genetics of the dog.* H.F. & G. Witherby, London.

Further Reading

In addition to the publications cited in the References the following are recommended to those readers seeking to delve deeper into genetics *per se* or into particular species.

General Books

Hutt, F.B. & Rasmusen, B.A. (1982) *Animal Genetics* 2nd edn. John Wiley & Sons, New York.
Lerner, I.M. (1958) *The genetic basis of selection*. John Wiley & Sons, New York.
Lush, J.L. (1945) *Animal breeding plans*. 3rd edn. Iowa State College Press. Ames.
Mason, I.L. (Ed.) (1984) *Evolution of domestic animals*. Longman, London.
Pirchner, F. (1969) *Population genetics in animal breeding*. W.H. Freeman, San Francisco.
Strickberger, M.W. (1968) *Genetics*. Macmillan, New York.
Van Vleck, L.D., Pollak, E.J. & Oltenacu, E.A.B. (1987) *Genetics for the animal sciences*. W.H. Freeman, New York.
Wright, S. (1968) *Evolution and genetics of populations. Vol 1. Genetic & biometric foundations*. University of Chicago Press, Chicago.

SPECIES BOOKS

Cattle

Cundiff, L.V. & Gregory, K.E. (1977) *Beef cattle breeding*. USDA, ARS AGR 101, Beltsville, Maryland.

Hinks, C.J.M. (1983) *Breeding dairy cattle*. Farming Press, Ipswich.
Johannson, I. (1961) *Genetic aspects of dairy cattle breeding*. University of Illinois Press, Illinois.
Schmidt, G.H. & Van Vleck, L.D. (1974) *Principles of dairy science*. W.H. Freeman, San Francisco.

Dog

Hutt, F.B. (1979) *Genetics for dog breeders*. W.H. Freeman, San Francisco.
Robinson, R. (1982) *Genetics for dog breeders*. Pergamon Press, Oxford.

Man

Cavalli-Sforza, L.L. & Bodmer, W.F. (1971) *The genetics of human populations*. W.H. Freeman. San Francisco.
Fraser Roberts, J.A. & Pembrey, M.E. (1978) *An introduction to medical genetics*. 7th edn. Oxford Press, Oxford.
Sutton, H.E. (1980) *An introduction to human genetics* 3rd edn. Saunders College, Philadelphia.

Pig

There appear to be no books in English devoted to pig genetics/breeding but the following have some material on the subject.

Devendra, C. & Fuller, M.F. (1979) *Pig production in the tropics*. Oxford University Press, Oxford.
Pond, W.G. & Maner, J.H. (1974) *Swine production in temperate and tropical environments*. W.H. Freman & Co., San Francisco.

Sheep

Land, R.B. & Robinson, D.W. (1984) *The genetics of reproduction in sheep*. Butterworths, London.

Owen, J.B. (1971) *Performance recording in sheep*. Tech. Comm. 20. CAB, Edinburgh.

Ryder, M.L & Stephenson, S.K. (1968) *Wool Growth*. Academic Press, New York.

Tomes, G.J., Robertson, D.E. & Lightfoot. (eds) (1976) *Sheep Breeding*. West Australian Inst. Technology, Muresk.

Wickham, G.A. & McDonald, M.F. (eds) (1982) *Sheep Production: Vol 1. Breeding & Reproduction*. NZ. Inst. Agric Sci., Wellington.

Journals

Numerous scientific journals in various languages are published throughout the world and many of them have information of value to animal breeders. It is not feasible to list all such journals but a sample of useful ones are listed below. Country of origin and number of issues per year are given. Papers do not necessarily stem solely from the country of origin. Journals marked * may include papers in languages other than English.

Acta Agriculturae Scandinavica (Scandinavia–4).

Animal Breeding Abstracts (Britain–12). This journal abstracts numerous papers of a genetic nature concerned with animals. It abstracts from over 1000 journals published in numerous languages. It also publishes (at intervals) very useful review articles. Essential reading for animal breeders hoping to keep abreast of their field.

Animal Production (Britain–6).

Australian Journal of Experimental Agriculture and Animal Husbandry. (Australia–6).

*Canadian Journal of Animal Science** (Canada–6).

Journal of Animal Science (USA–12).

Journal of Dairy Science (USA–12).

*Livestock Production Science** (Europe–12).

New Zealand Journal of Agricultural Research (New Zealand–4).

*World Review of Animal Production** (UN–4).

Index